THANK GOD FOR STEM

LIVING BY FAITH AND BY SIGHT

George Acquaah

George Acquaah Publishing

Published in the United States of America

ISBN 978-1-958518-14-4 (SC)

George Acquaah Publishing
21 Brigadier Ct,
Gettysburg, PN 17325
www.stellarliterary.com

Ordering Information and Rights Permission:

Quantity sales. Special discounts might be available on quantity purchases by corporations, associations, and others. For details, contact the publisher at the address above.

For Book Rights Adaptation and other Rights Permission. Call us at toll-free 1-888-945-8513 or send us an email at admin@stellarliteray.com.

CONTENTS

My name is Dr. George Acquaah, and this is "Thank God for STEM: Living by Faith and by Sight."

STEM is the acronym for Science, Technology, Engineering, and Mathematics. STEM is shorthand for the world of technological advances that inform our daily lives in 2022 .

This book seeks to help believers understand how they can embrace their faith in God and the abundance God provides through the application of scientific knowledge. I will argue that knowing how to balance religion and science is critical to the well-being of individuals, families, and our world.

CHAPTER 1

THANK GOD FOR STEM

The past two years have been especially contentious ones in our world. We have witnessed battles among neighbors, between nations, against disease, and over what we can believe. Today I ask you, "How might we usher in 2022 so that we can enjoy life *every* day with thanksgiving?" "How can we bring greater peace in our daily affairs and amongst one another?"

Yours, LORD, is the greatness and
the power and the glory and
the majesty and the splendor,
for everything in heaven and earth
is yours.

1Chronicles 29: 11

I believe the answer to both questions is, "By embracing and accepting the provisions He has made available to us."

As First Chronicles makes clear, there is nothing outside the reach of God. God dwells in our capacity to think, observe, build, and create. As a result, He has made Science, Technology, Engineering, and Mathematics (STEM) available to us. It is through our devotion and these gifts-these miraculous tools from God that we study His creation and learn the ways He benefits humankind .

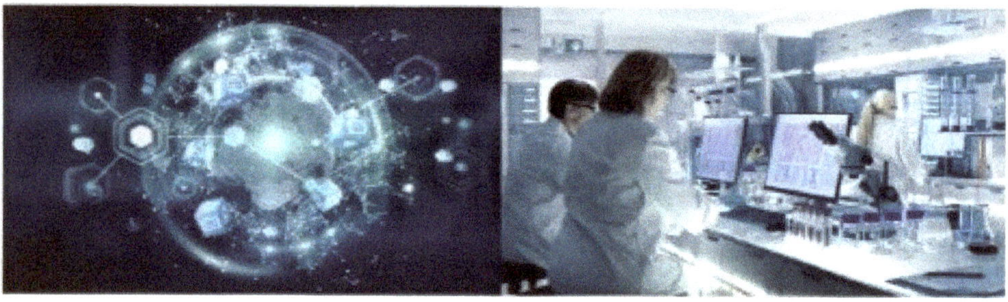

I have long been concerned about the misrepresentation of science in society and the perception that it is subversive of faith or Christianity. As a Christian who is a scientist, I have come to understand that I can be both without feeling dubious. Science, Technology, and God are compatible. It is a matter of proper perspective and making an effort to understand how to live by faith and by sight.

God is at the center of it all, for He made it all He is all-knowing; we are· not. It is unwise to define and package Him (God in a box, if you will) by our minimal knowledge about who He is, and presume to know His purpose for His Creation, completely. To place science, technology, engineering, and mathematics outside the purview of God is to diminish God and all the wonders He has set before us. Although I planned this project long before COVID-19 interrupted our lives, the pandemic lent urgency to its realization and my mission to help believers embrace the God-given power of science.

The subtitle of this book could very well have been "Living by faith and common sense" but I chose the current one for various reasons.

First, how to "live by faith and sight" will stop believers who know their Bible in their tracks and cause them to want to know what I mean by the audacious statement that seems to fly in the face of Biblical teaching (For we walk by faith and not by sight - 2Corinthians 5:7).

Living by faith and walking by faith are essentially the same concepts, but I chose "living" because it clarifies that I am not challenging the Bible. I am a devout Christian, and I am a scientist who has spent over four decades studying plants and their cultivation. My study reinforces my awe for the beauties and wonders of nature and science and affirms my faith in God.

Plant science, like any science, demands credible evidence and proof. Proof is based on what can be observed, tested, and corroborated. I have a responsibility to my field and humanity to contribute thorough, verifiable work. I am obligated to offer findings that meet the tests of the world and the laboratory.

But blessed are your eyes because they see, and your ears because they hear.

Matthew 13:16

During the pandemic, it was dismaying to witness how information and guidelines from the scientific community were dismissed in preference for ignorance and political opinions, which exacerbated the situation.

More disturbing was the fact that some religious leaders were willing to put their congregations at risk, based on what I believe was lack of knowledge and understanding of biblical truth. It is likely some Christians literally perished because of a lack of knowledge, as the Bible states, for they felt going to church or not forsaking the assembly of the brethren was the Biblical thing to do, against the advice of scientific experts to stay home.

On the other hand, it was encouraging that some Christian leaders, including those who are lesser known as well as some who are more visible in society, for the size of their following stepped up to the plate to give timely and godly counsel to believers. I even heard one big-name pastor warn against tempting God by defying scientific guidelines. He urged believers to exercise common sense, which I believe was the Biblical thing to do at that time.

HOLY
BIBLE

HOLY BIBLE

Christianity is not anti-knowledge, as some think. It is credited with founding some of the best educational institutions in the world at all levels of education, elementary through university. These institutions, especially the higher educational ones, were established to study theology and the liberal arts. These institutions have emerged to become world scientific research centers (e.g., Harvard, Duke, Princeton, Oxford). From the beginning of our most esteemed schools, the founders recognized the need for critical thinking reasoning and a spiritual foundation.

Human beings are thinking and reasoning beings. Christianity is not anti-intellectual for it is with the mind as well as the heart that we serve the Lord. Now, with the fingerprints of STEM all over every aspect of society, it is time to regard faith and reasoning as compatible and see science as a God given tool for the betterment of society.

The events associated with the COVID-19 pandemic confirmed that many Christian do not know how to handle science and faith together in daily living. I hope this effort will be helpful to many believers. I do not intend this book as a deep theological or scientific thesis, but rather a simple and general discussion of factors that hinder Christians from living life to the max, unencumbered by traditions that infringe on the true understanding and practice of faith as followers of Christ.

My goal is to inform and start a conversation, especially among people of faith, to acknowledge STEM's tremendous and indispensable role in modern society. Even though I focus on Christianity, I suspect other God-based religions like Islam and Judaism, may have their own struggles with how to live in our modern information and technology-based society.

I aim to benefit people from all religions and faiths. Still, because I am most conversant with the Christian faith, I will keep it simple by making all my references to the Bible, the authoritative book on Christianity.

I ask that you view the ideas presented here in the context of Scripture, and hopefully by doing so, you will realize, as I have, that STEM fits into the grand scheme of God's plan for the world. Christians and people of other God-based faiths can embrace and value STEM as a gift from God to help us live for Him.

The society in which biblical events unfolded was largely agrarian. Although the principles and concepts of the Bible are timeless if adequately understood and applied, the teachings and examples reflect the realities of those days. When we read the Bible today, we must take the information into context. We meet Scripture in the circumstance of our daily technological lives.

I am not debating whether or not the Bible should be taken literally. I am suggesting that we come to scripture in our imperfection and with our capacity to choose how the Word of God will shape our lives and our conditions today. STEM is part of our 21st century lives.

EVERYDAY LIVING 2022

Where the senses fail us, reason must step in.

Galileo

Galileo Galilei

I believe the Biblical rebuke that people perish for lack of knowledge is applicable in what I am trying to achieve with this book. People of faith should be willing to engage STEM, and the way to do so is to understand these fields reasonably.

I cringe when I hear pastors and other religious leaders use their pulpits to embarrass themselves and cast the faith in a negative light by misrepresenting science and scientists. Science-bashing by religious leaders is age-old, ask Galileo what the church at his time did to him.

I do not think it is necessary to believe that the same God who has given us our senses, reason, and intelligence wished us to abandon their use, giving us by some other means the information that we could gain through them.

- Galileo

I believe it is essential for everyone to study, not only the Bible for themselves, but also pursue other sources of information that they need to function in our modern society. To function well today, we need as many tools as possible. Contrary to what I often hear some religious leaders say.

Our society is influenced by three fundamental "forces" that are rooted in science, religion, and politics. We must understand them all to function at our best as social beings in a complex world.

Traditionally, when a religious leader stands before the faithful, we expect to hear from the throne of God, not human wisdom and opinions, something that is a fairly common occurrence today.

Some leaders have given in to politics and have turned their pulpits into platforms to promote and advance the political positions they fancy or deem advantageous to them. Some shamelessly allow themselves to be used by politicians and special interest groups.

Some leaders don't know what they are saying when discussing certain subjects because they have not studied them sufficiently or effectively. The Bible warns us of false prophets or leaders (I call them misleaders).

The Bible is a powerful tool or guide for Christian living. It can be an instrument for mass social transformation for good in the right hands. But it can be a weapon of mass destruction of humanity in the wrong ones. I am not talking only about religious events like the infamous Jonestown, Guyana, massacre of 1978, where a California-based cult perished through a mass murder - suicide orchestrated by their leader. I am talking about people being taught fundamental Biblical doctrines as faith, a practice I believe is often misunderstood and misapplied in Christianity.

We hear about "prosperity teachers" and "faith healers" who have massive followings. These preachers are often talked about with concern, even in the church, indicating that something is amiss in their messaging.

Divine prosperity and healing are Biblical concepts, if properly taught and applied. In this era of "mega churches" and "mega pastors," it is even more critical to ensure that leaders who are privileged to be entrusted with a massive following are extra careful about their messages. For false preaching can have an unprecedented and destructive effect on the lives of believers.

You don't have to take formal classes to understand STEM. Here you'll find an introductory overview of these fields, the rules that govern their operations, their applications in society, the concerns of the community with their deployment, and the basic information that would equip the believer to participate intelligently in discussions of these fields.

Alexander Pope once said, "A little knowledge is a dangerous thing." Still, I have deliberately provided just enough information about STEM and these forces, so as to whet the appetite of the reader, and encourage one to learn more about the various subjects or topics broached. Find the truth for yourself and own it. If you are depending on the Bible's account of nature to understand it, you will not get very far, for there is very little detail on the subject.

This work is meant to be readily accessible to the ordinary person without a doctorate in theology or science. After all, one does not need to be credentialed to believe in God.

On the other hand, STEM is different because it can be very technical. Consequently, I have introduced the reader to these fields in very basic and non-threatening ways to enable all to understand the intersection of faith and science that I seek to highlight. Christians do not have to feel dubious about living in a technological world. We can know what to embrace or reject without confidence rooted in God's word and not be deprived of religious traditions, politics, and social activism

The information here derives from my study and knowledge of the Bible, my personal life and experiences as a follower of Christ and servant of God, my professional background as a scientist and educator, and my general observations of nature and modern society.

Over the years, I have listened to sermons from preachers and ministry leaders and presentations and discussions by scientists on various TV and radio shows and documentaries. Some of the information I share comes from these sources.

Occasionally, I mention names of personalities who are widely known to society, e.g., Einstein and Darwin, or popular informational programs like the "cosmos," "How the Universe Works," and Star Talk."These are programs which enlighten me on cosmology, which I have never studied in my formal education. On purpose, though, I have refrained from mentioning individuals by name, be they politicians, preachers, academicians, or other professionals, unless doing so helps to make a point.

I may oversimplify some concepts, especially the scientific ones, in my desire to make this accessible to all users, but hopefully not to the degree of misinformation or blatant errors.I stand corrected. One of my goals is to encourage users' active pursuit of information. Consequently, I deliberately provide just enough information to make my case, hoping to entice you to follow up and obtain further information for a deeper understanding of the concepts as desired. I refrain from quoting Scripture, verbatim, but just how I understand it. After all, there are numerous translations of the Bible targeting various audiences.

I did not spend hours in the library to search for information, as I would if I were writing a formal book. As I worked on this project, I sometimes paused for what I call "just-in-time reference" to understand further or clarify some topics, concepts, and terms so that the information presented would be as accurate as possible.

Most of these quick references were via the Internet and included popular sites.

Most of the technical knowledge comes from my educational training and the various textbooks I have written on biotechnology, genetics, and plant breeding. I have several Christian pieces of literature in circulation as well.

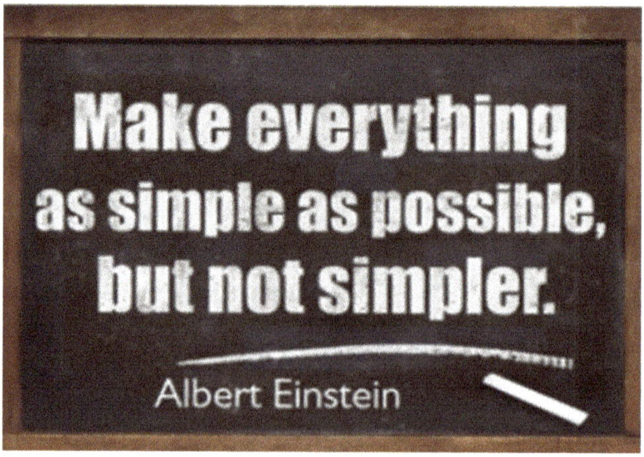

PRINCIPLES OF
PLANT GENETICS
AND BREEDING
24-7
LIVING LIFE TO THE MAX
MADE IN AFRICA
MADE IT IN AMERICA
UNDERSTANDING
BIOTECHNOLOGY
AN INTEGRATED and CYBER-BASED APPROACH
George Acquaah
HORTICULTURE
GEORGE ACQUAAH
PRINCIPLES OF
CROP PRODUCTION
Theory, Techniques, and Technology
GEORGE ACQUAAH
Practical Protein
Electrophoresis
for Genetic Research
GEORGE ACQUAAH, PhD

A quick way to access information of all kinds, including scientific, religious, and political, is the Internet, our window into the world. But be careful what you find when you search the Internet through this portal.

Popular sentiment today is "Don't believe everything you read on the Internet." This warning is accurate. There are unintentional untruths and blatant lies on this social platform. Coming from all walks of life, some people deliberately misinform people for personal or political advantage.

What is critical, in my opinion, is understanding the *essence* of scripture, holding it in one's heart, and acting on it. It is of little use to have the Bible in hand and not the heart.

I do not have a lock on facts and information, and I acknowledge, as a scientist, that new knowledge continues to be discovered or created. I don't want people, especially Christians, to hold on to the status quo for the sake of it and be deprived of living and enjoying life to the max, 24 - 7, according to the purpose of God. Some things are true, whether you believe or accept them, or not. They will impact or affect you, whether you know or acknowledge them, or not. Seeking knowledge about your world is Biblical and essential for living.

I will demonstrate the pervasiveness of STEM in society, through the eyes of an average person in a modern society. I continue by discussing nature and how it works, then discuss what science, faith, and politics are about, and outline how practitioners conduct their business. Before considering the intersection of science, faith, and politics in society, I will compare and contrast science and faith the intersection where the rubber meets the road. It is with the mind that we serve the Lord, says the Bible. I will discuss how the mind works and the games it can play on us. Truth must be our guide in all we do. What is truth? An understanding of this concept is at the crux of this work.

I will argue for STEM and its role and application in society. In doing so, I will discuss popular views and concepts prevalent in society on science, religion, and politics. This point cannot be emphasized enough: I do not intend to make this a comprehensive answer to questions on any complex issue. I provide enough information for one to comprehend the nature of science, religion, and politics, and be knowledgeable enough to intelligently participate in civic and cultural discussions. Whenever appropriate, I state my personal positions on the subject.

Even though, in my opinion, I believe that I make clear and strong arguments for my positions, I understand that some people will not hold the same views on the issues. That is okay, as long as these counter-views are grounded in facts and reliable information.

Coming up is Chapter Two: The Dependence of Modern Society on STEM.

CHAPTER 2

DEPENDENCE OF MODERN SOCIETY ON STEM

In this chapter, I demonstrate modern society's dependence on STEM through the eyes of an average person in a developed country. However, the example applies to varying extents in people's lives in many developing countries as well. It is not meant to be an exhaustive description but a gist of what could transpire in the daily life of an average adult.

It is summertime; the temperatures are in the upper 90s and sometimes even lower 100s. The heat is sweltering; the humidity is high, making it uncomfortable to sleep. This is a common scenario in many parts of the world during summer.

Fortunately, you have a roof over your head that could be rigged with the latest technological wizardry. Your room may be equipped with a climate control unit (air conditioner), with options to modify the temperature and humidity for a comfortable sleep and rest.

You may have gone to bed, sleeping in a well-crafted bed frame with a beautifully decorated headboard to please the eye. The bed may have a simple and basic mattress or one with options to regulate its temperature and even vibrate to massage your aching body. It may even have the option to adjust its configuration to conform to the desired architecture for extra comfort. Your mattress, most likely, is covered with bedsheets, accompanied by pillows and a bedspread for additional comfort in bed.

The bedroom is well-lit by light fixtures hung from the ceiling or a unit on the nightstand. You control the light by flipping a switch or clapping it on or off. You may even have the luxury of controlling its intensity and how long it stays on.

Sometimes, you use colored bulbs for certain effects. All these are courtesy of electricity. Hopefully, you had a sound sleep and good rest for the night. You need to wake up at a certain time to get ready for an event or task for the day (e.g., exercise, catch a flight, prepare for work, go to church, etc.). You used an alarm clock or the alarm function on your cell phone. You may be a person of faith and have the custom of reading Scripture and praying to start your day. So, you turn to the old reliable Bible on the nightstand or under the pillow.

Alternatively, in this high-tech era, you may just reach out to the cell phone, access the Internet, and voila, you have the Bible in various translations. You could also access an audio version and include some music for your quiet time.

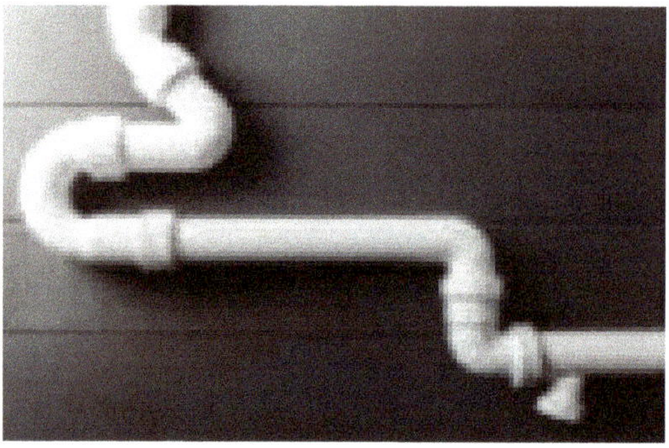

Afterward, you get to the bathroom to perform other essential daily rituals. You use the restroom and flush the toilet after cleaning with bathroom tissue. Then you brush your teeth and sanitize your mouth, using your choice of a simple manual toothbrush or a powered one. It is time to hit the shower. Here, too, you can select the water temperature and access a variety of cleaning utensils, soaps, and lotions. You may need to shave various body parts using your choice of manual and powered devices. To smell good, you wind up your cleaning routine with splashes and dabs of oils, creams, perfumes, etc. Note that you did not have to go to the river to fetch water for any purpose. You simply turned on the faucet, and from the pipes came clean and potable water. Your waste materials disappear through pipes to a waste treatment plant to be processed appropriately.

It is time to put on some clothes. You reach into the closet to pick up clothes appropriate for the season and occasion. If you do your own laundry, you use a washer and dryer, followed by pressing the items into shape with a pressing iron. Ladies may press and style their hair or simply cover up their natural hair with artificial hair (wig). Some go to the drawing board to make up for what God got wrong with an assortment of creams, pencils, and brushes, to cover up blemishes or enhance the original skin.

After cleaning and dressing up the exterior, it is time to take care of the stomach. You reach for the fridge or freezer for breakfast items that have been safely packaged and stored at the proper temperature (e.g., waffles, sausages, eggs, etc.). A glass of juice or milk may be your preference. You may prefer a bowl of cereal and a slice of bread toasted in a toaster. The grocery store offers amazing options of items to choose from. When appropriate, your food is pasteurized to kill harmful microbes (germs). After the meal, you may manually wash the dishes with cleaning utensils and detergent or load them into a dishwasher.

If you have health challenges, you may have a special daily routine of popping pills, as they say. For those who do not need to leave home, it may be time to relax in a favorite recliner or sofa to watch television, read, listen to music, or relax. For many, it is time to hit the road for work or an event. Depending on the distance from home, a form of vehicular transportation is needed. Depending on one's means, the sky is the limit as to the choices in terms of luxury and convenience available. The vehicles are powered by fuel and move on well-designed and constructed roads that are provided and maintained by the appropriate governmental unit.

The workplace is furnished with furniture, equipment, and other gadgets to enable workers to perform their tasks. Similarly, educational facilities are properly equipped for student instruction.

Our churches and other religious establishments are furnished adequately for the faithful to worship as desired, using technology to enhance worship with musical instruments. Buildings, whether residential or commercial, are carefully designed for safety and utility.

When the task or event is over, you return home to wind down for the day. Home is more than just a roof over the head. Here, too, the sky is the limit as to what is available, depending on one's means. While at home or even outside, one may communicate with people via various means, including cell phones. Modern phones are loaded with technological wizardry for communicating through social media outlets of all kinds.

Unfortunately, every now and then, adversities happen in life. Those of the health variety may be simple or very traumatic. If you are a parent, you know that a child may have a fever at any time. If you are a person of faith, you are taught to call on God first through prayer. But more often than not, if the fever is above 100 degrees Fahrenheit, the parent will reach into the medicine cabinet for a pain reliever or fever reducer. Occasionally, the health threat may be physical, such as occurs in a fall in the house or at the playground, and in the event of a vehicular accident. This calls for conveyance to a hospital or medical facility, often via an ambulance, where medical doctors and other staff are better equipped to provide appropriate care. You may go to the gym to exercise for a while, watch TV, and have a nightcap before turning in. The daily cycle starts all over again.

In the foregoing, I have attempted to provide a gist of a day in the life of an adult. While not exhaustive of what could occur, I believe the evidence is overwhelming to assert that without STEM, we would be living in the Dark Ages. We should be thankful to God for STEM. As indicated in the opening paragraph, the examples cited pertain more to an individual living in a developed economy. Needless to say, even in affluent countries like the US, there are pockets of severe lack where some of these amenities would not be applicable. Similarly, there are pockets of significant affluence even in developing economies like Ghana's. The causes of such inequities and disparities are many, not the least of which are politics, culture, knowledge, resources, and belief systems.

Let us participate in an exercise that would give further credence to my earlier assertion. I urge the reader to close their eyes for a moment and open them. If you are indoors, everything you see beside yourself has STEM written all over it! This includes pets. Unless they are wild species, all the dogs and cats are domesticated versions of their relatives in the wild. It took breeding, using scientific principles and concepts borrowed from Charles Darwin, who discovered the Theory of Evolution by Natural Selection.

In fact, the only thing that you don't see but is critical to life is air. Few people wake up in the morning and think about the air. We take this critical life-sustaining factor for granted. We think about the air when it is fouled up by accidents like burnt food, foul odors from rotten items, or insufficient amounts of air. Under such circumstances, we seek to let fresh air come in by opening doors and windows, for example.

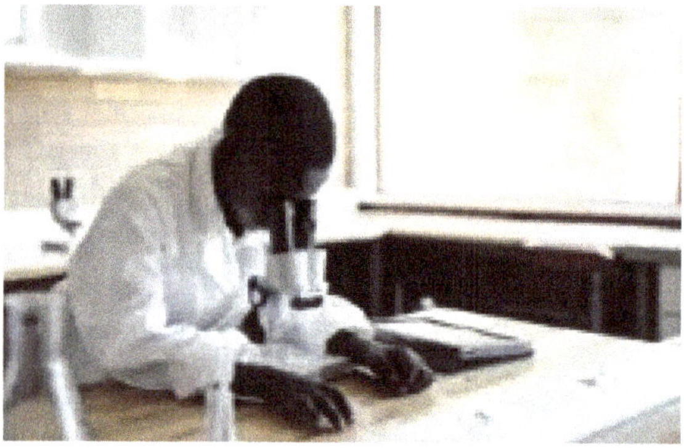

Of course, the other critical factor needed for life to occur and be sustained is water. You can see it when you tum on the tap or reach for a drink from the fridge. Bottled water may be on the table, but this water is portable and conveniently accessible via pipe or containers because of various processes that owe their existence to STEM. In fact, sophisticated technologies allow us to treat and recycle wastewater into domestic water sources. Astronauts treat and purify their liquid wastes for drinking while in space aboard the International Space Station. Purifying seawater is a significant source of drinking water for countries like Israel. All the things you immediately see around you when indoors, including your clothes that make life livable, are courtesy of STEM in one shape or form. Whereas air is one factor in the myriad that makes life possible and comfortable for humans, life would come to a screeching halt without it. In terms of life, the air is the equalizer, for it is freely available to all people, regardless of geographic or socioeconomic circumstance. Where there is life, there is air. We can do without water for days, but while without air there is no life, there is no quality living without STEM in our modern society. The discovery of vaccines (along with sophisticated medical equipment) is helping to significantly bring the COVID-I9 pandemic under control, thereby saving lives. Historically, all disease epidemics of our modern era have been contained or controlled through STEM and not through supernatural means.

It is impossible and, in fact, unnecessary to list all the ways STEM impacts society and the numerous technologies created to make these happen. One more example is worth mentioning. The

Biblical command by Jesus for his followers to proselytize the world is now made more accessible by technology - TV, radio, Internet, satellite, and social media technologies. Some Christians struggle with availing themselves of medicine, even though they would go to the hospital and receive other medical care to get well. Why single out medicine from the countless list of the benefits or products from STEM you routinely enjoy? The problem can be blamed on the false teaching of some religious leaders, especially the so called "faith healers" and Christian leaders (misleaders) who lack a proper understanding of Scripture. I believe in miraculous healing. But the preponderance of the evidence today points to God healing through His provisions He has made possible through STEM practitioners.

Consequently, we should thank God for ALL healing, with or without medical intervention. I hear it often said that God heals and doctors take credit. This is neither fair nor accurate. God works in partnership with humans to get things done on Earth.

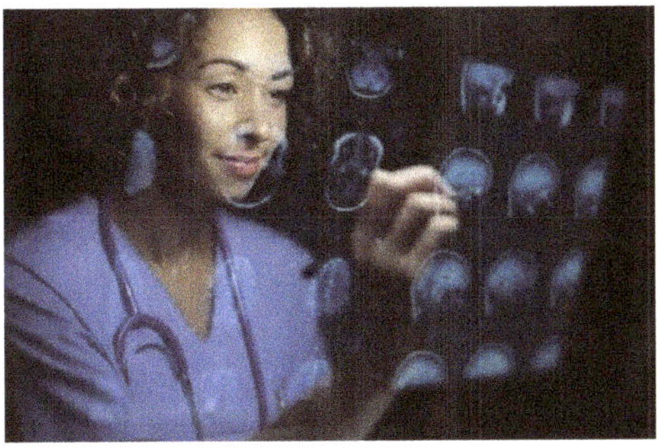

Because Jesus is called "*our healer*," and most of His miracles had to do with healing the sick, some Christians feel that accepting the modern miracles of medicine would make God obsolete in our lives. If we can do it, we don't need God, they think. On the contrary, if we were created in the image of God, it is to be expected that humans would do God-like things. It is all part of His plan for creation, and so we should embrace and thank God for STEM.

Let me end by saying that all these amenities of life described above are products of a vibrant economy. STEM is the basis of all industries that support the national economy.

It provides jobs and a better quality of life for the citizenry. If life in society is so dependent on STEM, it must be part of God's plan to help humanity carry out His charge to us to dominate and care for the earth. Of course, human beings drive STEM. We should then also be thankful to God for scientists and other practitioners of STEM. Herein lies the dilemma for some people of faith. Accepting STEM would appear to be giving glory to humans instead of to God, who has stated that He does not share His glory with any man. But that is not the case at all. After all, God created humans in His image and likeness. It is no accident then that humans are creative like Him. The apple, they say, does not fall far from the tree. If you are hungry, not knowing where your next meal is going to come from, and all of a sudden someone knocks on your door with food, would you accept the food or slam the door in the face of the giver? No, you would receive it and express profound thanks to your "angel of mercy. "Then, you would retire into your house and thank God for His timely provision of sustenance, through a human being. It is perfectly fine to give thanks and receive thanks; appreciate others and receive appreciation, as long as you realize and acknowledge that it is all God, and give Him thanks and praise.

In fact, the Bible teaches that believers are God's agents through whom He does His work on Earth. Yes, in the Bible, He sent ravens to feed someone, and Jesus paid his taxes with money obtained from the mouth of a fish. But how many times, if at all, over the last 1, 000 years, did such miracles happen again to people?

Yet, acts of kindness and provisions in times of need are routinely met through people all over the world on a daily basis. I am not suggesting that miracles of those Biblical proportions are a thing of the past. Absolutely not! God can meet needs in any way He pleases, with or without human involvement.

As people of faith, our role is to obey, and do what we know and are able to do, and leave the rest up to Him. We must not limit God by our limited minds.

It is clear to me that this is how God has chosen to routinely deliver his people through adversity today. Scientists developed vaccines and medicines to combat disease outbreaks in modern history. Let us thank God for STEM and all its enablers. God could have just intervened in the COVID-19 pandemic via supernatural means to wipe out the virus, but He has not done so. Why? I don't know. He is sovereign and can do as He pleases. But knowing what I know, I can confidently pray to God to provide wisdom and knowledge to scientists to develop cures and antidotes as needed to address any medical eventuality. I can also pray for decision-makers, politicians, and other operatives in the chain of discovery, production, and implementation of products, to act in the people's interest and not selfishly. Then again, I can also volunteer to assist in any way within my means. All these can be done to the glory of God. As indicated in the introductory chapter, the influence of STEM will only get more pervasive in society, not less. The sooner Christians get on board with STEM, the better it will be for all This is not a bandwagon that appears for a season. We are in this for the long haul.

In the next chapter, I will discuss the nature of the STEM disciplines. This introduces science disciplines in an easy-to-understand way to the uninitiated.

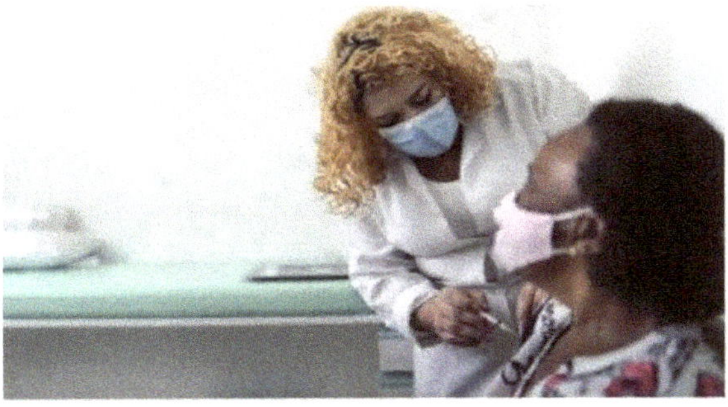

CHAPTER 3

THE FORCES THAT GOVERN OUR UNIVERSE

In the introductory chapter of this book, I mentioned three "forces" that govern our lives as social beings on earth, namely, science, religion, and politics. In this chapter, I delve into Scripture to show how I arrived at this assertion. Christians believe that God created the universe. According to the Bible, He did so over a period of six days. The account of creation is recorded in only two chapters (1-2) of the book of Genesis of the Bible. There are certain critical aspects of the creation story worth noting for future discussion in this project. First, it unfolded in response to voice commands given by God. "Let there be...," and there was whatever He desired and commanded. If the process were like an artist painting a portrait on a canvas, you would say His canvas was blank and dark, as the creation account goes. Now, could this have something to do with the ubiquitous dark matter and energy that scientists are pursuing as the next frontier in understanding nature? Just a thought.

God proceeded to populate the dark space with things over six days as follows: Day 1 - Light, Day 2 - Atmosphere or firmament, Day 3 - Dry ground and plants, Day 4 - Sun, moon, and stars, Day 5 - Birds and sea creatures, and on Day 6 - Land animals and humans. On Day 7- He rested or created the Day of Rest, as some suggest. We use the term "nature" to refer to the phenomena of the physical world and life in general that constituted the products of creation. However, humans and their activities are usually discussed separately. The Bible is illequipped to discuss nature in any appreciable detail. Later biblical accounts of life on Earth, like the story of Noah's Ark, tend to emphasize large animals. But through science, we know of the existence of billions of microbial flora and fauna (microbial plant and animal life).

The second feature of creation to note is how humans came into existence. Unlike all other phenomena and things that were spoken into existence, God created man in His own image and likeness, fashioning him out of the dust of the earth. He then breathed the breath of life into him, and he became a living being, from whose rib woman was later formed. What is significant to note about this unique process is that it endowed humans with creative powers, for they were made in the likeness of God the creator. The stupendous achievements of humans over the years should not be surprising, for, as they say, the apple does not fall far from the tree. Like father, like son. Adam and Eve, the first couple, were placed in the Garden of Eden, the Biblical Shangri La, so to speak, where they had the best of all they needed. The third notable feature of creation is that God gave humans dominion over all the earth and charged us to subdue it, be fruitful, multiply, and fill it.

Unfortunately, He did not tell us how to do that. How can you dominate something about which you are ignorant? By having dominion over creation, humans were placed in a superior role to exercise control and rule over all other life on Earth. Implicit in this command is responsibility and accountability to God, who gave us the assignment. It is expected that humans would do a good job and not run the creation enterprise into bankruptcy, if you will.

To subdue the earth requires that humans acquire a mastery of plant and animal life and the environment in which life is sustained.

Outside the Garden of Eden, humans learned how to cultivate plants and raise animals for food and other needs (the invention of agriculture).

Another feature of creation to note is that humans were granted free will or choice in their lives. They were not robots controlled and manipulated by God to do His bidding. They were free to exercise this option in life as they desired. In fact, they were free to disobey God's commands, something they eventually did when they ate the so-called "forbidden fruit." This exercise of choice was aided and abetted by a spirit being we refer to as Satan or the devil.

Choice has consequences, as we say, and what ensued after their disobedience was the mother of all consequences because it dramatically altered God's original plan for humanity. Humans became aware of right and wrong as sin first entered creation. The offending couple was swiftly banished from the Garden of Eden into the real world, if you will, to eke out a living. Choice, a God given right, continues to date to be a problem for humans, and is a source of great contention in the church and general society. Choice is so religiously and politically charged a term that it has become an ugly word in modern society. But God is the one who built choice into creation, so it must be good, even though it can be abused.

This brief background of the creation story is necessary to reveal the natural and human-made forces that control our universe. I choose to classify them as follows:

a. Physical forces

Physical forces are built into nature by God and control how the universe functions without human intervention. The physical laws of nature are universally applicable. As I will discuss later, researchers have revealed the four fundamental forces that control the universe as gravity, electromagnetism, the strong nuclear force, and the weak nuclear force. More details are forthcoming on these technical matters in future chapters. But suffice it to say that discoveries of these forces and laws that undergird them unleashed several global economy-boosting revolutions that have transformed our modern society today. For example, Newton's Laws of Motion Of Laws of classical mechanics) describe the relationship between the motion of an object and the forces acting on it, and stimulated the industrial revolution (steam locomotives, factories, and machine tools were designed and used). Newtonian science also is applied in aircraft and spacecraft design and operation.

b. Chemical and biological phenomena

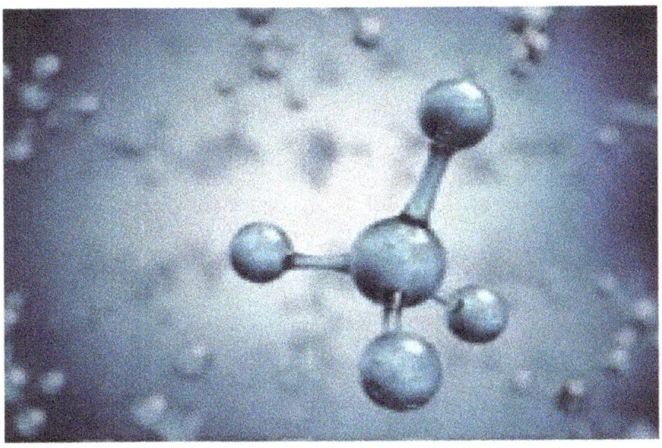

The physical forces operate in the heavens and on Earth. Life on Earth, both plants and animals, are controlled by complex natural biochemical processes (e.g., photosynthesis is, heredity, reproduction, respiration, etc.) that are all God ordained. Of course, chemicals occur in the planetary system and the general outer space. Similarly, by studying and unlocking these phenomena, researchers have unleashed biological revolutions that have transformed modern society (e.g., The Green Revolution that transformed classical food production and the more recent Biotech Revolution, which entails techniques that use living organisms or substances from such to produce or alter products). Biotechnology enables us to advance agricultural production and productivity beyond classical approaches (e.g., growing plants on artificial medium in the lab and engineering new varieties using radical techniques).

c. Spiritual forces

Spiritual forces exist in nature as invisible beings created by God. God himself is spirit, the ultimate good spirit. At some time in creation, He created angelic beings, one of whom would eventually rebel against Him and be banished from or cast out of heaven, along with his supporters. These outcasts became Satan and his demonic (evil) spirits on Earth. Spirits influence the actions of humans for good or bad in society. This started right in the Garden of Eden, where Satan deceived Adam and Eve to disobey God. Cartoons on decision making by humans often indicate an angel with a scythe and a pitchfork on one shoulder (evil influence) and an angel with a halo on the other (good spirit). Also, a common expression in society is "The devil made me do it," acknowledging the influence of evil spirits or forces in our midst.

d. Human force

Created in the image of God, given authority over all creation, and granted free will (choice), humans are a force to reckon with and can signnificantly impact the world in a variety of ways, for better or for worse. Even though denied by a minority of people and mainly for political purposes, climate change is a real and present danger to the earth and humanity. The role of humans in causing this adverse effect with unfolding devastating consequences is in controvertible.

In the Garden of Eden, humans enjoyed a relationship with the supernatural being that is God. The awareness of an invisible power outside of themselves, who could intervene in the affairs of humans, followed them into the larger world even after they were unceremoniously kicked out of the Garden of Eden. With free will, humans later sought relationships with other spiritual beings besides God, their creator.

Now, I want to show how humans capitalized on our creativity and free will (choice) to exercise dominion over creation as God instructed us to, through the "three human made forces" alluded to in the introductory chapter. We invented RELIGION as a means of worshiping and maintaining relationships with supernatural forces in the spiritual realm. In addition to God the creator, we worshiped the creation sun, moon, rivers, mountains, etc.

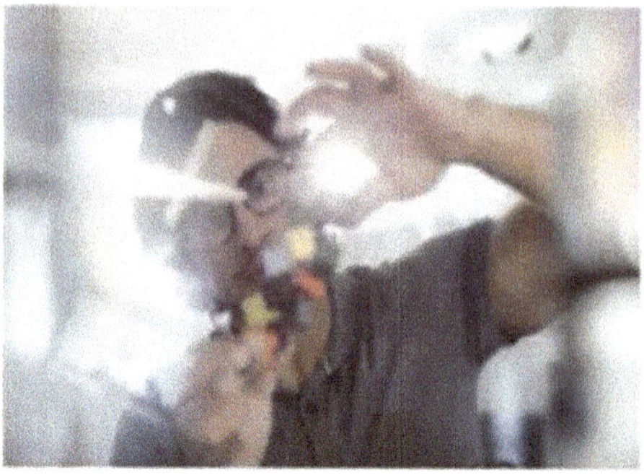

In order to live in harmony in a community, we invented the art of governance, whereby people agreed to abide by certain ways of conduct. Human communities today subscribe to various systems of law and order and maintain social structures to ensure the well-being of their members. These governance systems have evolved to become what we call POLITICS in our modern era. Even though we are endowed with creative powers, humans do not create anything from nothing as God did. We query nature to unlock the deep secrets of creation. With the knowledge gained, we invent things that make life on Earth better and more productive. Querying nature is an age old human pursuit. The modern way of querying nature is called SCIENCE, a universal method governed by a set of rules and codes of conduct. This is at the heart of discussions in this book.

Natural forces operate without the permission of humans (no respecter of persons). Gravity exists and operates, whether you know or believe it or not. The sun will come out according to its natural cycle without your permission. But how we harness the natural forces for our benefit in society depends on these three human-made "forces" - SCIENCE, RELIGION, and POLITICS, and how they interact. How they interact is influenced by how we exercise the God given right of CHOICE. The economic prosperity of a nation is driven by scientific discoveries and innovations that lead to inventions of technologies for engineering applications in our society. Therefore, it is critical that our modern society appreciates science and supports its advancement and application for the betterment of society. In this regard, the forces of religion and politics can be beneficial or detrimental to scientific progress, depending on the country. In a democratic society like America, religion and politics can significantly influence the advancement of science largely because of the power of choice.

History records a period called the Islamic Golden Age (8th-13th Century), during which science flourished under various caliphates. Have you ever heard of "Arabic numerals," which we use in modern mathematics? Among other things, algebra was invented in Baghdad. But as religious leaders gained political power, they declared the practice of science and manipulation of numbers to be secular undertakings that needed to be curtailed. Consequently, Islamic science began to lose its global leadership and dwindle. This hints at the fact that mixing religion and politics can be very dangerous for scientific advancement. I hope America will not go down this rabbit hole. Our politics is becoming increasingly divisive and filled with hate and decreasing room for reasoning together.

Unfortunately, religion that is supposed to have the power to change human behavior for good sometimes creeps into bed with politics in a manner that betrays the basic tenets of Christianity, which are - love for God, love for each other, the pursuit of justice for all, peace, grace and mercy towards one another.

In order to subdue nature, we must understand its structure and how it works. Science is the method by which we probe or interrogate nature. It can be very technical at times and needs to be explained at a level that the uninitiated can understand. Therefore, I will introduce this complex field of STEM in due course to facilitate discussions. Some people do not trust scientists because they don't understand how they do their work. Religion is an area where opinions abound due to interpretations and misinterpretations of Scripture. However, it is not hard to go back to the basics to find true religion that is not adulterated with human traditions. I intend to discuss how faith works or should work, according to the Bible and not denominations. The third human force, politics, is a beast. It can be used to manipulate society. I introduced the notion of truth in the introductory chapter of this project. What is truth? This notion is discussed in more detail in a future chapter. But for now, let it suffice to say that there are three basic kinds of truth: Science seeks and operates by objective truth; faith operates by personal truth; politics operates by convenient truth. If one understands the characteristics of the three forces, it becomes easier to engage them in society. As will be discussed in a future chapter, the three forces must meet at some point in society, and when that happens, politics is usually the arbiter. This is so because in a democracy, where the rule of law prevails and politicians preside over the affairs of a nation, it is not uncommon for religious folks and scientists to lobby lawmakers for what they need.

Going forward, the ensuing chapters will cover the following topics:

The nature of STEM disciplines

How nature works

How STEM works

How science works

Mind power and STEM

How faith works

Science vs. faith

How politics works

Science vs. politics

Choice, ethics, morality, and STEM

Living by faith, science, and politics in society

When STEM goes wrong

CHAPTER 4

THE NATURE OF STEM DISCIPLINES

Science, technology, engineering, and mathematics (STEM) were shown in previous chapters to drive a modern economy. This chapter discusses these disciplines in simple terms to allow the average person to understand enough to participate intelligently in discussions involving their application in society.

MATHEMATICS

According to experts who study the field and the complex entity that is the universe, mathematics and the universe are intimately linked. Some refer to mathematics as the language of the universe. "Mathematics is the language in which God has written the universe"(Galileo). Mathematics (simply called maths or math, depending on whether you are from Europe or America) is the study of numbers, shapes, and patterns (an abstract science of numbers, quantity, and space). The language of math is one of the abstract relationships based on patterns that practitioners use to invent artificial order from chaos. Mathematics and logic are bedfellows, and the patterns and relations they address are considered universal truths. Mathematics allows practitioners to make accurate predictions about the universe.

In order for scientists and researchers to arrive at objective truths, they abandon their natural senses - sight, taste, sound, touch, smell (which are subjective) - and instead use tools for measurements to reveal reality. Then they develop mathematical models of this reality that can be manipulated logically (math is about the logical extension of one point to another) to enable new discoveries about our world. Through STEM, researchers have tools that aid their quest to understand nature in various realms that are not accessible to the human experience (i.e., beyond the capacity of our senses). For example, the microscope (from the basic light microscope to the electron microscope in their various dizzying arrays of types and sophistication) allows researchers to explore our world at the tiny (microscopic) end of the organization of nature. On the other end of the spectrum (grand scale), the telescope (e.g., the Hubble telescope and the James Webb telescope) allows us to explore the universe on a vast scale. These technological marvels are essentially extensions of our very limited human eyes. The laws of physics operate in these two realms (minute and infinite in size). At the same time, mathematics allows us to traverse (with our minds) this size complexity spectrum in incremental steps without bias (but without the capacity to experience it in an immersive way).

Cosmologists believe that the universe is one huge mathematical structure for all intents and purposes. Further, all matter consists of particles with properties (e.g., charge and spin) that are purely mathematical. Even space may have properties like dimension, but ultimately it is a mathematical structure. Another way of putting it is that the physical universe is mathematics per se and not just a description by math. All structures that exist mathematically exist physically (mathematical existence = physical existence). So, where are humans in this scheme of things? We are described as "self-aware substructures" in the physical universe.

I heard it said on Dr. Neil deGrasse Tyson's Star Talk that the mathematics of the LASER (Light Amplification by Stimulated Emission of Radiation), a machine that makes an amplified beam consisting of of light, was figured out by Albert Einstein decades before the modern laser devices were invented. The 2016 movie Hidden Figures celebrated the outstanding contribution of an African American mathematician, Dr. Katherine Johnson (human computer), to the success of NASA's space program. She calculated the precise trajectories that enabled the Apollo 11 spacecraft to land on the moon in 1969. In the movie, astronaut John Glenn requested the NASA team leader obtain confirmation from Katherine before he embarked on his historic flight. In fact, research programs like those at NASA employ mathematicians to provide the theoretical basis of their research. It has been said that, without mathematics, modern civilizations would not have emerged.

Mathematics is fundamental to STEM. There are various branches of the study of mathematics. Pure mathematics is the study of mathematics for its sake (no specific application in mind), while applied mathematics is the application of mathematics in other STEM disciplines. In addition to the study of quantity, structure, space, and change (i.e., arithmetic, algebra, geometry, and analysis), mathematics is linked to fields like logic (especially in computer science, philosophy, semantics). New subspecialties of applied mathematics continue to be developed in higher education, including engineering math, mathematical physics, mathematical biology, mathematical finance, etc.

Additionally, because of the pervasiveness of math in our world, it is one of the required courses or subjects for students at all levels of education (i.e., General Ed. requirement). Mathematics helps develop aptitudes in creativity, critical thinking, problem-solving, and reasoning, among other things. Unfortunately, many students approach mathematics with a defeatist attitude that it is difficult. Hence, they are their own prophets of doom at the end of the class when they flunk the course (self fulfilling prophecy).

I will close this introduction to mathematics with some fun things that can be observed in nature that are undergirded by mathematics. Composite flowers (occur in clusters as opposed to simple flowers), e.g., buttercup, delphiniums, com, marigold, have whorls of petals that occur in numbers described by the so-called Fibonacci series (1 12 3 5 8 13, etc.). The series is computed by summing up the latter two numbers to give the next number in the sequence. The sequence is also observable in hurricanes, spiral galaxies, shells of some crustaceans, and others.

The other common mathematical factor found in nature is called the Golden Ratio (or divine proportion, as it is sometimes called) and is represented by phi (ϕ), with a value of approximately 1.618 (it is an irrational number - goes on infinitely). It is obtained by dividing a line into two parts, the longer part (a) and the smaller part (b). The golden ratio is observed when $(a + b)/a = a/b$ (if the proportion is in the Golden Ratio, the quotient will be equal to approximately 1.618). The rule is applied in creating artwork that draws the viewer's attention to important elements. Various body parts exhibit this ratio (e.g., forearm to hand). The Mona Lisa is said to be one of such paintings that demonstrates the concept. The ratio is featured in the architecture of buildings.

SCIENCE

Now, let us look at science. Science is the pursuit and application of knowledge and understanding of the natural and social world, following a systematic methodology based on evidence (Science Council). It is a way of discovering the universe or nature and understanding how it works. This quest for knowledge follows standards and a systematic process called the scientific method (discussed in detail later). Science as a field is very expansive and diversified. It may be classified into four categories or branches: math and logic (also called the formal sciences), biological sciences, physical sciences, and social sciences. Biological (also called life sciences) and physical sciences are also described as natural sciences. They focus on studying the nature of living things and natural phenomena associated with the universe. As the name implies, social sciences focus on studying human behavior and societies in general. Natural and social sciences are also referred to as empirical sciences because the knowledge they generate is based on observable phenomena that are subject to verification by peers. Empirical evidence is simply information that is acquired by observation or experimentation.

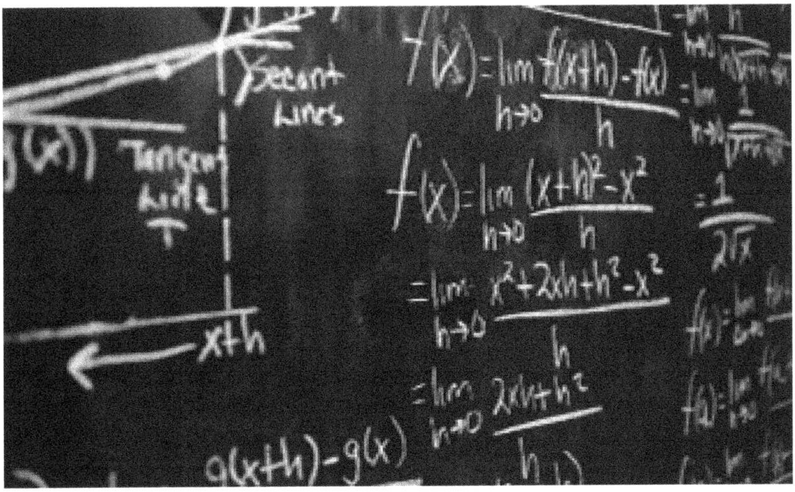

Mathematics is related to science and is considered by some to be the purest form of science. Science assumes that our universe is ordered and subject to precise mathematical laws. The most fundamental of the branches of science is physics. Physics is expressed in terms of mathematical equations. It also undergirds chemistry and biology. I heard Dr. Michio Kaku of NYU summarize the scientific branches this way: All biology can be explained by chemistry, while all chemistry can be explained by physics. As said previously, physics can be explained by mathematics. It is because of this interrelatedness of the branches of science that, in academia, courses in programs of study are strategically sequenced in their offerings to allow the student to progress smoothly by being equipped with the appropriate prerequisites at each stage.

The four categories of science consist of numerous sub-branches or disciplines that are the basis of academic studies. Physics, one of the oldest academic areas of study, focuses on matter and energy and their interactions with motion and other forces.

Biology is the study of life, while chemistry is the study of matter and how it changes. The discipline of geology is involved with studying the earth, while astronomy focuses on understanding the universe. Other science disciplines are agriculture, zoology, genetics, botany, oceanography, biotechnology, bioinformatics (which combines biology and computer science), and many more. The Italian physicist, astronomer, and engineer, Galileo Galilei, is sometimes referred to as the father of modern physics, observational astronomy, science, and the scientific method (evidence based scientific research).

Science and scientists often butt heads with the religious establishment. Galileo, as previously noted, defied the Pope and published some research findings that ran counter to church teaching. For this, he served years under house arrest. But nothing compares with the controversy stirred up by Charles Darwin's seminal publication, On the Origin of Species, which runs contrary to the church's teaching of Creation as how the universe came into existence. I hope the discussions in this book will bring science and religion to the table for practitioners to reason together for the common good of humanity.

There are various types of scientists who focus on the different sectors of society to improve human life. They work in agriculture, medicine and health, meteorology, space and exploration, mining, drug discovery or medicinal chemistry, product safety, policymaking, economics and finance, and social study, to name a few, helping to make human existence and life better. They work collaboratively with others in the STEM quartet to solve real-world issues for the benefit of society (applied science). However, sometimes scientists pursue their quest for knowledge for its sake, without any application in mind. This type of pursuit is called basic science (versus applied science). It might take a longtime before someone finds a useful application of a theoretical discovery. Basic scientists produce the raw material for innovation. Unfortunately, some in society do not have the patience to wait for basic science to yield dividends and are quick to question why they should be allowed to continue on public funds. This kind of attitude and mentality towards science is dangerous and stifles creativity and innovation. Policymakers should be educated on this matter and would do well to yield to the counsel of competent science advisors. Basic science must be optimally funded in a sustained manner if America is to keep its global competitive edge over other nations.

ENGINEERING

Engineering is defined by The American Engineers' Council for Professional Development (ECPD, formally ABET) as: "The creative application of scientific principles to design or develop structures, machines, apparatus, or manufacturing processes, or works utilizing them singly or in combination; or to construct or operate the same with full cognizance of their design, or to forecast their behavior under specific operating conditions; all as respects an intended function, economics of operation and safety to life and property." Simply put, engineers apply the principles of science and mathematics to solve real-world problems facing society by creating or inventing, designing, and building things that solve technical problems to address societal and consumer needs. The things they build include roads, bridges, vehicles for transportation (cars, planes, boats, etc.), buildings, tools, computers, and processes.

Before things are designed and built, all kinds of mathematical calculations and the applications of scientific principles and concepts are considered. You may recall I previously mentioned that the mathematics of lasers was laid out by Einstein decades ago. But it was not until 1960 that Maiman developed the first laser device. Now, we have laser technology in printers, survey equipment, medical equipment, industrial cutting tools, barcode readers at checkout stations in the grocery store, and many more applications.

Inventions are not always preplanned pursuits. One of the most beloved household appliances, the microwave oven, was invented purely by accident by an observant scientist, Percy Spencer, in 1945. A radar engineer, he observed while working near the magnetrons (high-powered vacuum tubes inside radars) that a peanut butter candy bar in his pocket began to melt. Putting two and two together, he later brought popcorn into the vicinity of the magnetrons, and lo and behold, popped corn was flying everywhere. He had discovered that microwaves could be used to cook food in seconds. His employer, Raytheon Company, patented the discovery and designed and produced the first commercial microwave oven two years later. Just like computers, the original model was huge and clunky. It was not until it was reduced to counter-top size that the product's popularity took off, making it an essential household appliance today.

There are a number of TV programs that celebrate modern engineering achievements, such as the skyscraper, sonic flight, subways, bullet trains, spacecraft, the International Space Station, long bridges, dams, locks, etc. These are huge structures, but some engineering marvels are miniature devices. Now, the microphone headset used by public speakers is barely visible to the audience. Some wristwatches can rival desktop computers in what they are capable of doing. What about man's favorite device, the TV remote control that has contributed to the couch potato lifestyle of many people? I would be remiss if I did not point out that some of the most amazing engineering achievements belong to the ancient world. The mighty pyramids of Giza in Egypt still amaze modern-day engineers and architects. One of the most talked-about ancient achievements is the wheel, which is so simple and yet so versatile in application. Have you ever heard of or used the expression, "Do not re-invent the wheel"?

Engineering is a very popular academic field of study for students, making it a competitive one to get into. Major branches of the field include Mechanical, Chemical, Electrical, Civil, Geochemical, and Management, with a plethora of subcategories (e.g., Aerospace, Biomedical, Computer, Software, etc.).

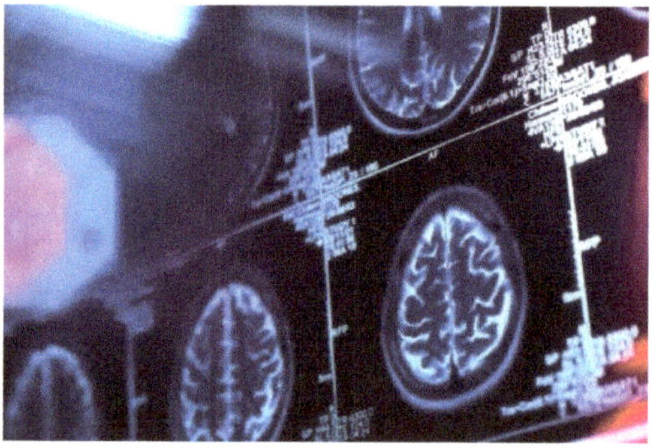

TECHNOLOGY

One key difference between science and technology is that the principles and laws of science, once established, remain pretty stable and change infrequently. On the other hand, technology changes rapidly, sometimes so much so that it is hard to keep abreast.

In short, the things engineers create, design, and build constitute technology. This includes skills, processes, and methods used by ordinary people and professionals to solve specific problems. Technology is used in diverse ways, such as to produce goods and services, scientific research to make new discoveries, medical applications to save lives and assist in space exploration. Some believe that our modern world would not be in existence without technology. This fact is clearly presented in Chapter Three of this book. In fact, scientists use technology to answer very complex questions and discover new knowledge faster.

We have come a long way from stone tools and stone wheels to pneumatic tires, mechanical hammers, and electric drills.

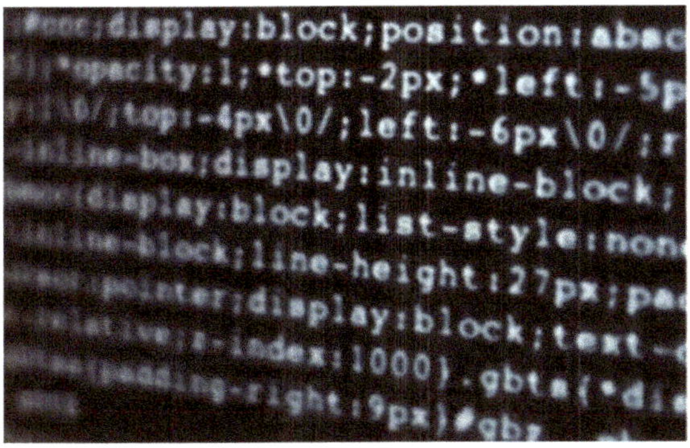

Technology may be grouped into categories according to applications such as medical, construction, manufacturing, energy and power, transportation, communication, agriculture, and biotechnology. Technology evolves rather quickly. One time, I was in a meeting with architects and engineers working on a major academic building when one person commented at one point that "no one repairs things these days. "This was in relation to some electronic equipment that was being discussed. It is cheaper and easier to replace than repair some of these devices or equipment. Your current cell phone was practically obsolete when you walked out of the store, for new models are already in the queue, waiting to be strategically time-released. Computers and many other early generations of devices have been miniaturized for agility. Devices are increasingly efficient and faster in execution and data sharing, mobility, remote connectivity, and automation. Technology has increased productivity in the workplace. But technology is not always hardware. Software engineers and computer programmers routinely design and release new versions of software for new versions of devices.

Computer-enabled technologies dominate our modern society. The Internet, a household innovation, has revolutionized communicating and sharing information, conducting business, and educating students globally. The COVID-19 pandemic compelled many institutions to migrate many of their programs online permanently. The Internet has spawned the phenomenon of social media. New technologies have transformed record keeping and retrieval, data analytics, and decision - making. These include artificial intelligence (AI), virtual reality and augmented reality, robotics, machine learning, blockchain, edge computing, cloud computing, cybersecurity, and the Internet of Things (IoT). America is back in space because of advanced technologies developed by the likes of SpaceX and NASA. Some of the most amazing technologies are applied in the medical field for diagnostics, surgery, and automation of processes (e.g., MRI, CAT SCAN, robotic surgery, incubators in NICU, X-rays, echocardiogram, etc.). Robots are now a familiar presence on the floors of manufacturing enterprises (assembly lines). In higher education, the lines between science, humanities, and arts are blurring because of the pervasiveness of technology in academic disciplines.

I hope this brief introduction to the STEM disciplines was helpful. Of the four, science is the primary one that will be talked about in future chapters. Scientific discoveries drive technology and engineering applications. Science is how we understand nature (the Book of Nature). Using fundamental discoveries of science, engineers invent technologies that facilitate more discoveries.

The way scientists go about their work, the discoveries they make, and the applications of the discoveries in developing technologies are of concern in society. The problem is that some people jump into discussions emotionally rather than with information. My goal is to provide information to assist people in intelligently participating in social debates and discussions about STEM. I cannot emphasize enough that STEM is here to stay, and therefore it behooves us all to understand and engage properly for the benefit of society.

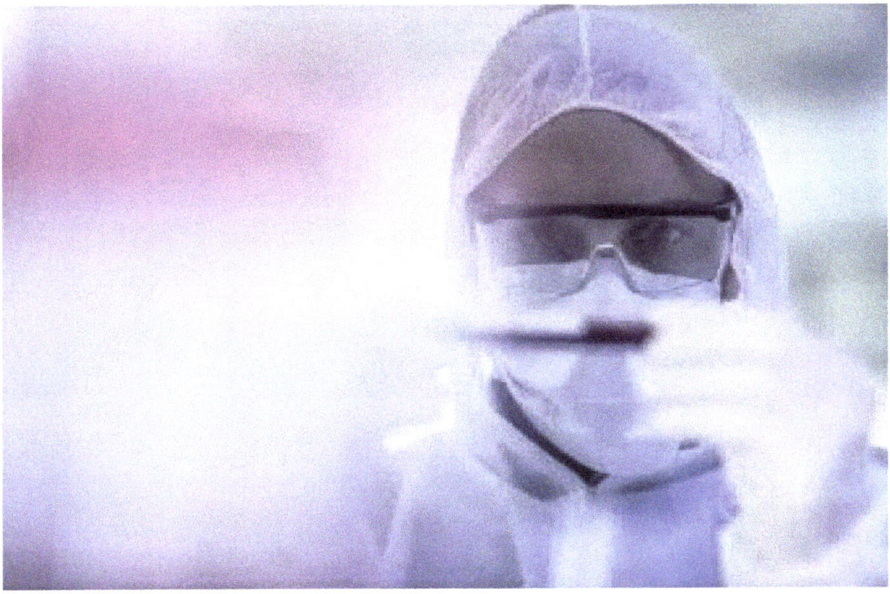

CHAPTER 5

HOW NATURE WORKS

Broadly speaking, the term nature refers to the natural, physical, or material universe. It can also refer to life in general and technically include humans, but humans and their creations are often considered apart from natural phenomena. The key message of this chapter is to emphasize that God is the author of nature. Dr. Hugh Ross of the Reasons Organization calls it the "Book of Nature." If that is so, then I believe it should be on par with the Bible in authenticity and authority. Both books contain divine truths. The difference is that God directly authored nature, while persons under His inspiration wrote the Bible. Unlike the Bible, in which some of its authors have rendered some accounts a little differently, the Book of Nature is free from human errors. Science, a human endeavor, is the method by which we elucidate and decipher the truths of nature. Consequently, understanding nature is a work in progress. Whenever we find apparent contradictions in either one or between the two books, it could be that we lack understanding of the issue or are misinterpreting one or both of them.

The significant thing about nature is that it is self-sustaining. The Christian religion teaches that God created the universe and that He is in control of His creation. But I sometimes wonder how He controls or manages it. Walking along the street, I see a leaf suddenly move as a slight wind blows in from nowhere. Then I think to myself, did God move it on purpose? After some time, I conclude, in my opinion, that God built into His creation mechanisms that are self-triggering, self-regulating, self-sustaining, and interconnected. Some natural laws and processes produce causes and effects in nature and cycles that ensure that order is maintained. Having done that, God now sits back and watches as nature takes its course. If a wind blows and encounters a leaf and stone, it might have the power to move the leaf to another location while the stone is left undisturbed. Several laws of nature were in play in the simple event (gravity, power of the wind, etc.). The leaf might be run over by a car and torn into pieces in the new location. Did God plan to do just that? I don't know. Of course, He can intervene at will to change the natural course of things in the universe. He is omniscient and omnipotent.

Starting from the top-down, the amazing planetary system shows various celestial bodies that move in their orbits with predictable regularity and precision. The solar system is self powered and regulated. Scientists can predict the occurrences of spectacular celestial events like solar and lunar eclipses with high accuracy. A solar eclipse occurs when the sun, moon, and Earth are aligned, such that a portion of the earth is engulfed in a shadow cast by the moon (i.e., the moon comes between the sun and Earth and partially blocks sunlight). A lunar eclipse occurs when the moon returns the favor and moves into the earth's shadow (i.e., the earth passes between the sun and moon; without sunlight, the moon becomes dark).

Let us narrow it down and consider something about Earth that most people do not know. Did you know that the earth's axis is tilted by about 235 degrees, unlike other planets in our solar system? So, what? That is why we have seasons on Earth winter, summer, spring, and fall! Different parts of the earth are oriented towards the sun at different times of the year. The summer season is warmer than winter (in each hemisphere) because the sun's rays hit the earth at a more direct angle during summer than winter, and also, the days are longer than the nights during summer. Of course, Christians will say that God deliberately created the earth to tilt on its axis. Scientists look at this odd situation in the solar system and wonder how the tilt came to be. The suggestion is that it was hit by something massive when the earth was young. It is worth noting that this axial tilt of the earth (obliquity as it is called) has varied between 22 2'33"and 24 30'16"over about five million years.

Now, down to earth, where the rubber meets the road. This is where humans engage nature most intimately. Nature is sustained through phenomena like natural cycles that replenish resources on Earth. Scientists have determined that all life on Earth is carbon-based. Carbon is the primary component (45-50%) of all dry biomass (organic material made from plants and animals). Without carbon, there would be no DNA, proteins, sugars, fats, lipids, or anything that makes up what we call life. It is one of six major elements of living organisms - carbon, hydrogen, oxygen, nitrogen, phosphorus, and sulfur - which we represent in high school science with the mnemonic acronym CHONPS. These major elements are maintained in nature through what scientists describe as biogeochemical processes, or natural cycles, whereby the elements are converted into different forms through consecutive processes that involve living and nonliving systems of the earth. When an organism dies, its body undergoes decomposition to release these basic elements into the soil and atmosphere. Plants absorb these nutrients for growth and development. Animals feed on the plants and obtain the nutrients for their growth and development. Then, eventually, they die, and the cycle starts all over again.

In addition to natural cycles, nature is filled with natural rhythms that guide all that we do. Our heart beats according to a natural rhythm, just like our breath. The doctor checks our pulse when we go for a routine checkup. This natural rhythm can be altered by voluntary activities like exercise or events like fear, stress, or disease. Natural rhythms that orchestrate our lives include the sun's rising and setting, by which we begin and end our daily activities. The daily temperature changes (daytime vs. nighttime) and seasonal temperature changes impact our lives. In the wild, some animals under take annual rituals of seasonal migration for survival.

These natural rhythms are predictable and can be enjoyable to humans just to behold (e.g., the spectacular fall colors and the blooming of certain flowers). Nature has a built-in internal clock called the circadian cycle. This 24 hour cycle is controlled by physiological processes in all living things plants, animals, bacteria, etc. This "body clock" tells our bodies as humans when to eat, sleep, rise, etc. Though internally generated, the cycle is susceptible to external stimuli like sunlight and temperature. You may be aware of the caution by experts to avoid artificial light from electronic gadgets during the night to avoid adverse impacts on the quality of sleep.

When you go to the restroom at night, you'd rather feel your way to the stool and hope you don't ram your toes into the bedpost, all because you don't want to tum on the light to interrupt your sleep cycle.

Nature has a way of maintaining balance. Individual organisms have a self regulating process (homeostasis) by which they maintain internal stability, adjusting to conditions that assure survivability. This condition of optimal functioning is achieved through the work of physiological processes that maintain a relatively stable equilibrium between interdependent elements in the body. Common examples now part of modern medical practice include blood pressure, blood glucose level, body temperature, and breathing patterns, which are standard checks a doctor performs on you during a routine medical appointment. When necessary, the doctor will order blood work to ascertain much finer biochemical information (e.g., sodium levels, calcium ions, potassium, etc.). Thank God for STEM, for an imbalance in any of these homeostatic systems can be detected and corrected with medical prescriptions (diet, exercise, drugs, meditation, rest, etc.). Without this natural mechanism, the organism or you will die.

On a macro level, nature maintains a balance in populations in an ecosystem (a biological community of organisms interacting among themselves and their physical environment). These two entities (biotic living, and abiotic-nonliving) are linked through nutrient cycles previously alluded to in this chapter, and energy flow. One of the basic natural processes we learn in science, even at the elementary level, is photosynthesis, the means by which energy enters the ecosystem from sunlight. Animals feed on plants and facilitate the movement of matter and energy through the system, which also includes the indispensable role of microbes (i.e., the ecosystem has producers, consumers, and decomposers called the food chain). Everything in nature works in a delicate balance. Ecosystems must maintain what scientists call ecological balance in order to thrive.

Nature has a way of keeping the proper load of life in an ecosystem. Every organism has a natural enemy. Natural enemies in nature operate as predators, parasites, or competitors. For example, birds and bears eat fish; lions eat gazelles; snakes eat mice, etc. This may be likened to natural checks and balances in the ecosystem. If snakes are removed, the population of mice will grow out of balance. The US has several major ecological challenges resulting from introducing an invasive species for one purpose or another. The Florida Everglades is overrun with Burmese pythons introduced as exotic pets. Owners discarded these pets in that environment when they no longer had use for them. Without any predators in America and with a high reproduction rate, the pythons have overwhelmed the Everglades. Starlings are a menace to aviation, sometimes flying into and disabling jet engines with catastrophic results. The story is that one hundred of these birds were deliberately released in the New York area by an overzealous William Shakespeare fan, who got the idea from the play Henry IV, in which the bird is featured. Without predators, starlings now dominate the skies in parts of New York!

The ecological balance can be significantly disturbed when free-thinking, innovative, intelligent, and creative humans are included in the ecosystem. Activities of modern life threaten the ecological balance to the detriment of the quality of life and even the survival of life on Earth in general. I will illustrate with a few examples. Our modern hightech agriculture has increased crop and animal productivity, but not without adverse consequences to the environment. Runoff of agricultural chemicals (fertilizers) into water bodies (lakes, streams) triggers the rapid growth of algae, causing oxygen levels to be depleted. Fish cannot be supported under such conditions (called eutrophication). In a different scenario, scientists introduced the toad and snail into bodies of water in the Philippines to control natural species like water hyacinth. However, without natural enemies to control the foreign species, the population of snails and toads grew out of proportion and became pests. The invention of plastics revolutionized our lives in more ways than one. But the improper disposal of synthetic products like plastics, which cannot decompose to fit into the natural nutrient cycles, is choking bodies of water and harming marine and other lives all over the globe. How about climate change, a real and present danger running amok in the world with devastating consequences?

The evidence is incontrovertible that humans are the culprit in this state of affairs. Yet, the will to take action to reverse this catastrophe is frustrated by the beast called politics.

In addition to human influence on natural ecosystems, nature has ways of disrupting its own balance, often catastrophically and spectacularly. Examples of such natural phenomena include terrestrial events like earthquakes, floods, droughts, volcanic eruptions, landslides, and extraterrestrial events like a visitation by asteroids. These natural phenomena and events can change the physical landscape by creating mountains, valleys, plains, etc. Vegetation and animal life can be destroyed and possibly cause the extinction of some species. According to scientists, dinosaur extinction is attribute dto an asteroid's devastating hit on the earth. Such sudden changes on Earth can create conditions where the existing organisms are forced to adapt to their new environment or perish. The new environment becomes suited to a new set of species (ecological succession). These natural disasters are called "Acts of God" in modern society because they arise spontaneously without human instigation. Insurance companies will not hold to your charge damage caused by such events when you file a claim.

There are biological phenomena in nature that are equally devastating as physical events. Diseases and pestilence are all part of nature. Locusts can be devastating to crops, just like drought, and were cited in the Bible. Sometimes, when we infringe upon the habitats of wild animals, we put ourselves in proximity to their diseases. We can also bring such problems into the human community by domesticating wild life for food and as pets.

In addition to these spectacular events, some of which can unfold right before our eyes, other major events take millions of years to occur. One such event is the Ice Age, in which the earth comes into the grips of colder global temperatures, causing ice sheets to persist or move (glaciation) across the earth's surface with various consequences. A popular movie by the same name, Ice Age, entertains kids today with some of the living creatures of those times (e.g., the wooly mammoth). The earth consists of layers. The outer shell (lithosphere) is dynamic, slowly but constantly moving with various consequences. Scientists have a theory called plate tectonics to explain such a phenomenon. The lithosphere is fractured into several major rocks (plates) moving apart (plate tectonics). These movements are implicated in violent events like volcanoes and earthquakes, resulting in modifying the earth's surface. If you look at a globe and use a little imagination, you can visualize how the seven continents of the earth, once upon a very, very long time, were part of one huge block (Pangea or supercontinent).

Built into nature are forces that sustain it. This includes gravity, electromagnetism, and nuclear forces. It is true scientific querying of nature that laws associated with these forces have been elucidated. Applying these laws has fueled various revolutions (e.g., industrial, green, and biotech) that have been drivers of our modern economy. The planetary system operates under various laws of nature.

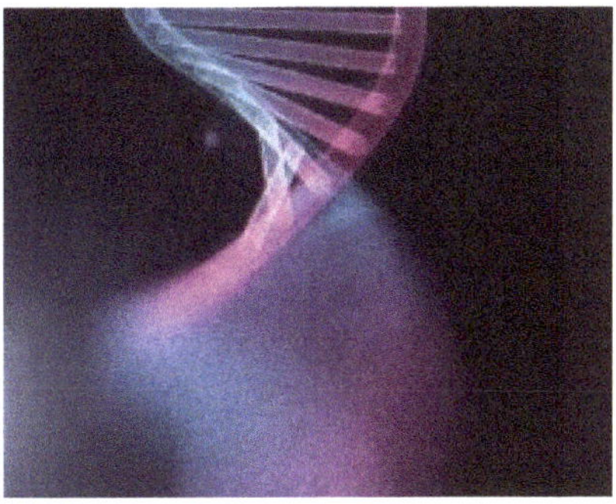

Before I close out this chapter, let us bring humans into the picture one more time. At creation, God instructed humans to be fruitful and multiply. This is possible because of two natural mechanisms heredity and reproduction. DNA is the hereditary material of humans. It ensures that each organism reproduces after its kind. Self replicating life has built in mechanisms for reproduction. DNA is susceptible to modification by natural processes, instigated by errors in internal processes (e.g., cell division) or external influences like radiation (e.g., ultraviolet rays from the sun). These factors cause abnormalities called mutations, which can modify the DNA and the hereditary mechanism to create new biodiversity. Mutations fuel evolution. Modern advances in genetics show that we are more than our genes. What we end up becoming depends on our behavior and environment. Called epigenetics, it is the study of how cells control gene activity without changing the DNA sequence (epigenetic changes are reversible).

In summary, God designed into nature various mechanisms and laws to enable it to operate in a self-sustaining manner. As reasoning, intelligent, and creative beings in nature, we like to probe and interrogate nature. With the information gained, we try to nudge it to our advantage. Also, because of the way nature works, it is best to understand it thoroughly so we would know how to protect ourselves against surprises and know how to "tame" it for our advantage. We achieve all of this thanks to STEM. Because we are limited in our knowledge and understanding of nature (but we continue to press on relentlessly), our innovations and applications do not always go well (see Chapter 13). Sometimes, we design vehicles (e.g., rockets) to defy gravity and fly into space, with varying success.

Air travel is now a routine mode of transportation. We launch communication and other satellites into higher orbits using technologies that defy gravity and other relevant natural laws. It is critical to point out that such human interventions do not alter natural laws or forces in any way whatsoever. But we thank God for STEM that enables us to counteract natural forces and laws to our advantage. Because we can place satellites in space, we have powerful technologies like GPS! How can we find our way to grandma's house without this modern technology? Sadly, sometimes, human ingenuity and innovation are no match for the forces of nature. Earthquakes can reduce our buildings to rubble in seconds; tornadoes and tsunamis can raze a town to the ground; our dykes and retaining walls can be overrun by the effects of hurricanes. Humans have a lot to learn from nature, so we can better subdue it to our advantage.

CHAPTER 6

HOW SCIENCE WORKS

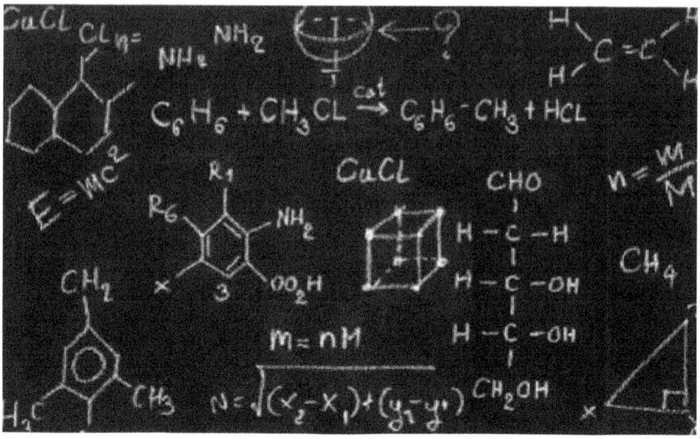

The primary purpose of science is to help us understand how nature and the society that dwells in it (technically also a part of nature) work and then apply that knowledge gained for the benefit of society. As previously indicated, engineering and technology depend on the discoveries made by science as building blocks. Scientists go about their business by following universally agreed-upon standards and systematic processes embodied in what is called the scientific process. This process is undergirded by certain values and codes of ethics that guide the conduct of all scientists. This way, discoveries they make are universally acceptable as truth, scientific that is, and applicable all over the globe. I should point out that some believe that science is value-neutral. We will explore this further in a future chapter.

Scientists are guided by their curiosity and creativity but, above all, the honest pursuit of objective truth. Being human, they are subject to the frailness in character that is common to us all and can therefore be flawed in how they pursue their work. Humans are plagued by cognitive bias (i.e., things you think are true but are not and can be shown to be untrue). Similarly, as scientists, we are susceptible to biases that are conditioned by the cultural, political, and religious influences in our environment. These biases can get in the way of how we pursue truth objectively. In their desire to be the first to discover something, a few scientists are known to have fudged data to say what they wanted and not what the reality was. Some of such lapses in judgment in the profession are due to the competitiveness of the academic process, whereby one's inability to contribute significantly to knowledge in a field could lead to denial of promotion or dismissal from the academy (publish or perish threat). Also, being the first to have discovered or solved a scientific problem is not only prestigious, but a patent of your discovery could make you and your organization very rich and famous. To be ethical, scientists must report authentic results and not withhold useful information from the public. You may have heard it sometimes said how some companies or commercial organizations withhold scientific research findings that expose flaws in their products because they fear economic loss.

Fortunately, because of the built-in fact-check mechanism in the peer review process, research results are not taken at face value by the scientific community. It will be fact-checked vigorously if it is too good to be true. Moreover, before a finding can be accepted as truth by the scientific community, it should have been independently verified and confirmed by others. It is more powerful when others use the same or even different methods yet arrive at the same conclusion. The preponderance of the evidence is the arbiter of scientific truth. The researcher must be rigorous in conducting their work and use sound methodologies and reliable measuring tools and techniques so that their findings will not be in dispute. It is common for colleagues to challenge or poke holes in others' findings by pointing out flaws in methodologies or the interpretation of data analysis.

The scientific process consists of several simple steps. The following is meant to introduce the uninitiated to this practice, and is not meant to be an exhaustive discussion of the subject.

a. **Observe** – Nature can be breathtaking in its splendor, majesty, and awe. It is only a matter of time before one would have questions about various aspects of it, including its existence and function. Also, as you observe society, questions will arise about the behavior of people or how the society functions.

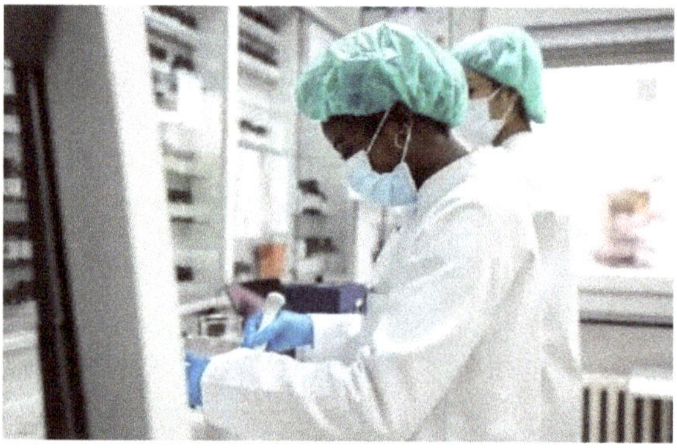

b. **Ask a question** – Ask a question, but note that science is not capable of answering “why?", only “how?”, "what ?", when?", "where?" questions. For a why question, one needs a belief system (religion). So, as I heard someone say, we leave that to prophets and priests. Religion is culture of faith; science is culture of doubt (Feynman). I will discuss more in a future chapter. What intrigues or puzzles you about nature or society that you want to know more about? This becomes the purpose for which you conduct a scientific inquiry. You are likely to find an answer only when you ask the right question. Not that a question may be stupid, but one can pose a question that is not appropriate or is fuzzy and cannot be researched. One of the all time great observations and questions in science is attributed to Isaac Newton. We often hear about how he observed an apple fall which led to the discovery of the force of gravity. But at the same time, he also observed the moon, which led to the great question; "If the apple falls, does the moon also fall?" The answer is yes! His investigation of the question led to calculus, which he invented to help him with his research. More importantly, the ensuing laws of motion that he discovered were responsible for the Industrial Revolution that changed the world. Read more details on this at your leisure. So, if you find yourself spinning your wheels in your research, it could be that you asked or are investigating the wrong question.

c. **Research the purpose** – The scientist may have a question or puzzle that other scientists have already addressed. Consequently, the scientist will first scour the existing literature (a literature survey or search) to see if the problem has already been addressed. If so, there are several options available to the researcher, including repeating the study with modification, if necessary. The more research outcomes that corroborate each other, the more compelling it is to accept the outcome as fact or truth. The original work may have been conducted with flaws in methodology, inconclusive outcomes, narrow scope, *etc.* Opportunities abound as to how to conduct research to contribute to existing knowledge or create new ones. This step in the scientific method acknowledges that scientists build upon knowledge, brick by brick, filling in the gaps as they go along to complete the puzzle eventually.

d. **Pose a hypothesis** – Once a question or problem has been identified, the researcher makes an educated guess (prediction) about its answer. This prediction is put in the form of a statement that can be tested in research. This step is very critical. It is framed contrary to the prediction. For example, suppose the purpose of the research is to determine the effect of nitrogen fertilizer on the yield of tomato plants (i.e., you suspect or predict fertilizer will increase yield). In that case,your prediction statement will be something like, "Application of nitrogen fertilizer will not increase the yield of tomato plants" (this is called the null hypothesis - conjectures that there is no difference between the yield of plants that received fertilizer and those that did not this is the culture of doubt I alluded to earlier). A hypothesis is scientific only if, through experimentation, it can be disproved (falsifiable).

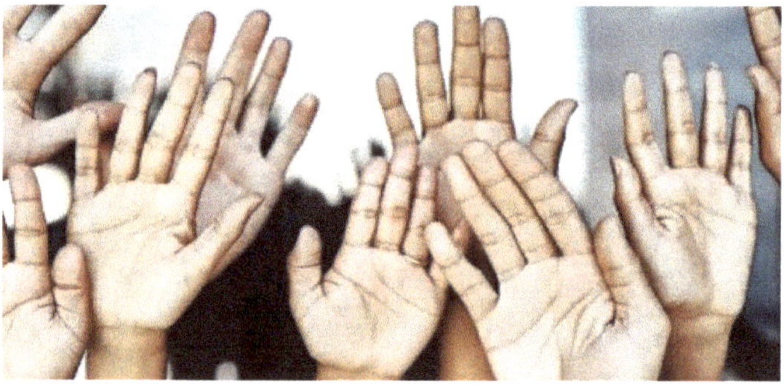

Research the hypothesis – Once a testable hypothesis is formulated, the next step is to design an experiment to confirm (accept) or disprove (reject) the null hypothesis. The design will include standard elements such as a control, replication, and randomization of treatments, variables, and sample size. We often hear it announced on TV or other news formats that a preliminary drug study included a small group (sample) and that researchers plan to include a larger group in future studies to improve the validity of the results. The scientific tradition is to disprove the null hypothesis because it is impossible to prove that something does not exist. However, you can disprove that something does not exist by discovering an example of it. So, in our fertilizer example, we disprove the null hypothesis of no difference by finding examples of plants that show an increase in yield.

There are two basic variables in research design - dependent and independent. A variable is a factor that can take on different values in the research. In our fertilizer example, fertilizer and yield are our two variables. Yield (dependent variable) depends on fertilizer to assume different values, while fertilizer (independent variable) does not depend on yield. In research, scientists manipulate the independent variable to elicit responses in the dependent variable. Instead of investigating one level of fertilizer, for example, we usually investigate multiple levels (e.g., 10, 20, 30 kg/acre). These levels of fertilizer are called treatments. Similarly, in a drug study, the researcher may investigate different drug doses in a treatment study. The control group will consist of plants that will receive no fertilizer to ensure that the researcher can determine whether the yield is caused by fertilizer and not any other factor. In human drug studies, the control will be a placebo (a substance with no therapeutic effect).

e. **Analyze the results** – The research design dictates how the data collected would be analyzed to reveal differences if they occur. There are standard research designs available to scientists, each with appropriate analysis methods. Again, it is common to hear it announced on the news that, for example, a drug study was conducted by the "double blind" design (i.e., neither the participants nor the researcher knew who received a particular treatment). This is a way in which scientists reduce bias in such experiments. Statistical methodologies are used in the analysis to test the hypothesis. Traditionally, the researcher tests a hypothesis not with absolute certainty but at a statistical level of significance (the probability of rejecting the null hypothesis when it is true). Ideally, the researcher wants to confirm reality; if there is treatment effect or difference, *we* want to find it. Statistical significance does not equate to the everyday meaning of big or important. For example, a 0.05 % level of statistical significance indicates that the researcher is willing to accept a 5% risk of concluding that a difference exists (fertilizer does increase yield) when in fact, it does not.

A statistical significance indicates that the effect we observed in our sample in the research was not by chance but occurred in the larger population from which our sample was drawn for the study. Finding statistical significance (probability or P-value)at a smaller value (e.g., P= 0.001 vs. 0.05)is desired.

f. **Make a conclusion** – From the evidence resulting from the data analysis, were you able to disprove your null hypothesis? That is, were you able to show that, indeed, the nitrogen fertilizer did increase yield? How do you explain the finding? How does your finding agree or disagree with previously reported ones you read about in the literature review? What recommendations are you willing to make in light of your findings?

It must be emphasized that one scientific finding does not constitute scientific truth. For the finding to be accepted as a scientific fact, the experiment must be duplicated by others - the more, the merrier, in this situation. If your critics confirm your results or use other methods to arrive at the same conclusion, all the better. Once established as fact, especially in laws of nature, it is not often that the verdict is overturned in the future, even though this is occasionally possible. My daughter, Bozuma, sent me this quote:

"If you are not a scientist, and you disagree with scientists about science, it's actually not a disagreement. You're wrong. Science is not truth. Science is finding the truth. When science changes its opinion, it didn't lie to you. It learned more."(Mohamad Safa).

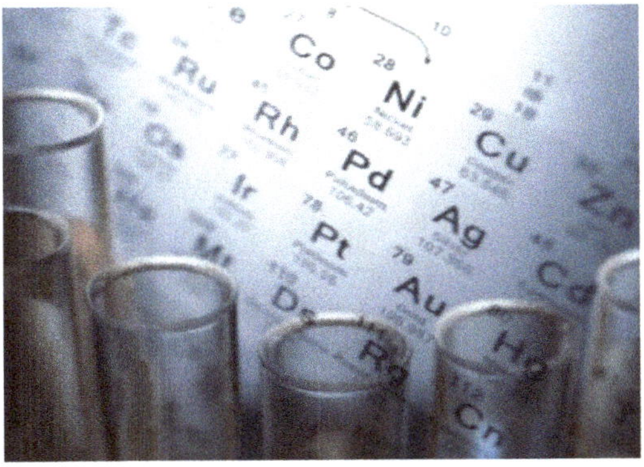

A principle designed to protect society from the abuse of human subjects in research is called the IRB (Institutional Review Board) approval. This mandate was instituted due to the unethical use of humans as subjects in the infamous Tuskegee Syphilis Study. Before a researcher conducts biomedical research or any research involving human subjects, the FDA mandates that formal approval must be obtained from an IRB. Universities and other research institutes have constituted their IRBs that review such research in accordance with FDA regulations for compliance before granting permission for the research to proceed. It should be heartening to society to know that the government is looking out for their interest as it pertains to research. In high-risk research like genetic engineering, scientists require safe facilities that meet approved standards (e.g., biocontained spaces, special ventilation, etc.) and approved disposal systems. If radioactive materials are stored and used on-site, special approval must be sought from the appropriate state agency with oversight of such matters.

Let me now share the selected ways scientists go about their research and practice. It is impossible to discuss how the numerous categories of scientists go about their business. I have selected a few that the ordinary person will readily identify with because they are usually in the news, for better or for worse. Food is critical for the survival of humanity, and so is health. Applied plant and animal scientists use scientific principles and concepts to nudge nature to the advantage of humans. In other words, we look at what exists in nature and try to improve upon it in ways that make life better for humans. As will become apparent, He procedures for improving plants and animals vary on a theme. If you understand one, you can understand the other. We are confronted with enormous biological diversity on a daily basis. A visit to the grocery store will reveal numerous varieties of produce. You might see over a dozen varieties of apples alone. The florist section would display a wide variety of flowers and greenery. Horticultural diversity is even more impressive in the landscape as we walk through the neighborhood. All this diversity results from the work of plant geneticists and breeders like me. Gene transfer is a routine technique we (as well as animal breeders) use in our work to create new varieties.

The most natural gene transfer method is crossing (hybridization). If a crop variety farmers are growing is deficient in a gene needed for better performance (e.g., disease resistance), we identify a source of the desired gene and transfer it to the cultivated variety through artificial crossing so that the improved variety can be whole and more productive. We follow the principles that emerged from the work of Charles Darwin in developing his theory of evolution to create a new product (plant variety). The Russian scientist, Vavilov, described breeding as evolution by the will of man (1935). The first step is to identify a source of heritable variation (e.g., a mutant with a novel genetic expression resulting from a mutation). Not all variation is heritable (genetic-based). Then, we need a mechanism for transferring the novel gene into another plant and reproducing it (e.g., crossing or recombining the two plants). Hybridization will lead to a segregating population as the two parents' genomes (the set of genes in each parent) are recombined and reshuffled into new cells. Finally, we need a method for searching for the products of the cells with the desired combination of parental genes (called selection). The natural process (evolution) uses natural selection that can take millions of years to create a new product. In modern breeding, we use artificial selection that allows us to develop a new variety injust about ten years or so. You might disagree with Darwin's theory of evolution, but when you sit down to enjoy a meal from modern agricultural produce and meat, give him some thanks for laying out the general steps in breeding. In other words, thank God for Charles Darwin.

g. **Domestication** is the process by which plants and animals from the wild are adapted for human use. Nature has equipped organisms in the wild with characteristics and mechanisms for survival. These characteristics are undesirable when the wild organisms are brought to live with humans in the same ecosystem and to be used for food or other purposes. For example, plants in the wild are equipped with morphological and chemical properties that ward off predators (e.g., bitter principles in fruits, thorns, thick fruit walls, tiny hard fruits for wide dispersal and domination of the environment, etc.). However, humans prefer sweet, juicy, succulent, and good-looking tomato fruits. Through selective breeding, plant breeders have stripped the wild tomato of its natural protection that gets in the way of its use in modern society. The price we pay for our actions is that we then have to protect domesticated plants through artificial protection like pesticides. Pesticides are polluting our environment and causing health challenges, so now we have to pay to clean up the environment. Wild rose plants have thorns for protection from being eaten by animals in the wild. But who would like to give or receive thorny roses on Valentine's Day in the name of love? And so, breeders have developed thornless roses, in different colors and sizes, among other traits.

The fundamental principle we use as breeders is summarized in this simple equation: $P = G + E + (G \times E)$, where P = phenotype (the physical appearance of the organism), G = genotype (the genetic makeup of the organism), E = environment (the environment in which the organism resides), and $G \times E$ = the interaction between the genotype and the environment. To make things simpler, we usually reduce P and G to represent specific traits and genes of interest (e.g., height, color,yield,etc.). The genetic information responsible for the physical trait that we see, is encoded by the DNA of the organism. Back to the equation, we say that if you don't like what you see (P), you have three basic options - change G (genes), E (environment), or both (G and E). If your tomato has a disease, you can change the production environment (e.g., by spraying a pesticide to kill or control the pest),or change the variety (use a disease- resistant variety).

If you opt for a pesticide, you must always apply it when you grow your tomato, in order to keep the disease under control. In other words, the pesticide is not a cure but a preventative action to control the disease. On the other hand, if you change the variety, the G, you are good for a long time. Plant breeders are about making genetic changes that are a permanent and heritable solution to a disease problem.

A cross can be made if the species is sexually reproductive. But what if the crop is asexually reproduced (does not use pollen or eggs)? Or, what if natural sources of the desired gene are not found in nature? Plant breeders have conventional and nonconventional techniques in their toolkits. As radical as some of these may appear, they occur in nature. The tools used are collectively called biotechnology, which may be broadly defined as the use of living organisms and biological systems to create modified or new organisms or products that are useful to humans. Genetic engineering is a biotechnology that focuses on manipulating the DNA of an organism, usually directly, in the lab via modern processes, to create individuals with new characteristics or functions, or new useful products. However, biotechnology is not confined to labs, and in fact some of the technologies are age-old. Fermented products, yogurt, cheeses, breads, soy sauce, etc. all result from the aid of microbes, and so do vaccines. In modern days, we use microbes to clean up some toxic sites.

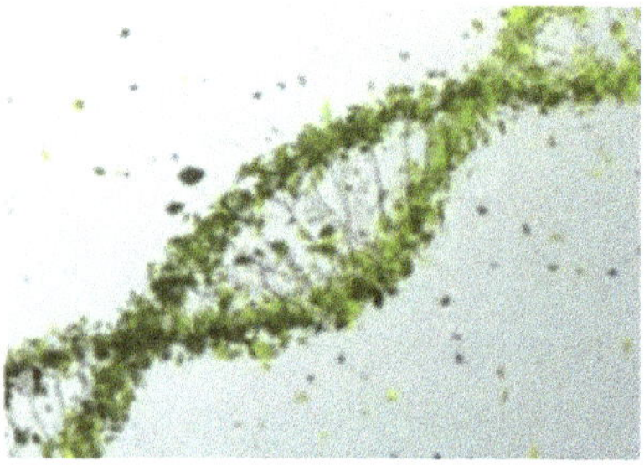

One specific and common technology used to generate variability for breeding is induced mutation (mutagenesis). Mutations occur routinely in nature (called spontaneous mutations). In news broadcasts, you may often hear that a virus (e.g., flu) may mutate, rendering the current vaccine ineffective. These are natural mutations. However, scientists can subject plants, whole or parts, to agents of artificial mutations (mutagens) that may be physical (e.g.,x-rays, gamma rays) or chemicals (e.g., EMS, colchicine) to cause physical or chemical changes in the DNA The consequence is a change in plant characteristics and performance. Hundreds of varieties have been developed using this technique.

Unfortunately, the Darwinian steps in breeding I just described are possible only if the two plants are reproductively compatible (can be crossed). So, we first search from close relatives for desired missing genes of interest before going to the wild relatives (very difficult to cross wild plants with domesticated ones). Sometimes, the gene of interest resides in what we call an alien source (cannot be naturally crossed with our domesticated plant). This calls for more radical tools like recombinant DNA (rDNA) technology (generally called genetic modification, GM), which has become controversial in our modern society. But truth be told, some of these radical biotechnologies were copied from nature (microbes like viruses and bacteria). For example, viruses lack nuclei and cannot multiply on their own. The only way a virus can survive and reproduce is by invading the host's genetic apparatus (in the nucleus) and hijacking it for multiplication.

This is how the COVID-19 virus survives, and that is why experts desire for us all to get vaccinated so that we deny the virus the means for reproducing.

The virus splices its DNA into the host's DNA. So, when the host cell multiplies, the viral material is also multiplied. How ingenious. In rDNA technology, scientists first insert the desired gene (DNA) into a specially engineered bacterium or virus (called a vector-the process is called cloning). A common bacterium used in gene cloning in plant research is the Agrobacterium tumifaciens (it causes tumors in plants like tomato). The tumor-inducing DNA is removed by scientists so the bacterium cannot cause tumors anymore. Then the desired gene (DNA) is inserted into its place. When the bacterium infects a susceptible plant, it reaches out to the tumor-causing DNA section and transfers it into the host plant as force of habit, to start doing damage. Only this time, it has been tricked by scientists, for they have replaced the tumor DNA with the desired DNA, so it now starts to do good instead. What a switcheroo.

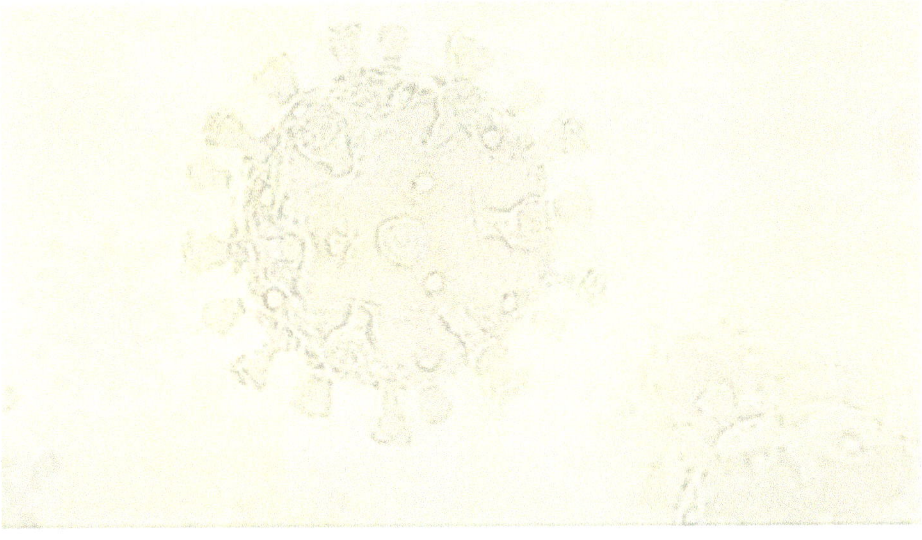

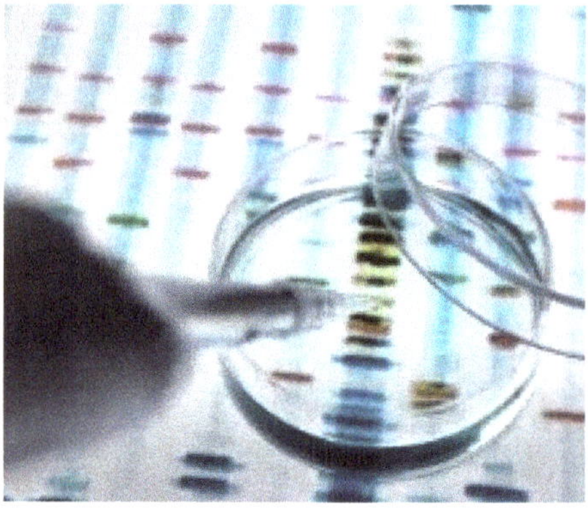

The recombinant DNA (rDNA) technology is used in the two most dominant applications in modern plant breeding the development of Bt crops and herbicide tolerant crops. The European corn borer is very devastating to corn production. Scientists discovered that a soil borne bacterium, Bacillus thuringiensis (Bt), produces natural toxins that can control certain insect pests like the corn borer. Using the rDNA technology, breeders have transferred the bacterial gene that produces toxins to insects into the corn plant. This allows Bt corn to produce endogenous toxins to kill the corn borer without the need for spraying pesticides during corn production!

In the case of herbicide tolerance, Monsanto's glyphosate herbicide (marketed as Roundup®) is popular in corn production. It works by blocking the EPSPS enzyme in plants. Scientists cloned a gene from a soil bacterium that produces a glyphosate-tolerant form of EPSPS and inserted it into corn. When the GM crop is planted, the farmer may just spray the entire field with Roundup, killing only the unprotected weeds while leaving the protected corn plants unharmed. The corn varieties produced by this method are said to be genetically modified (GMO).

One advantage plant geneticists have over human and animal geneticists is our latitude in manipulating our research subjects. We can cross plant parents almost at will to create new variations and study the inheritance of traits, or create new varieties of plants and do so without fanfare. You cannot just cross two humans for the sake of research studies! We can genetically engineer plants in ways that human and other animal geneticists cannot for ethical and other social concerns. Plant geneticists can manipulate the germline (reproductive system via pollen and eggs) to create variation that can be generationally heritable. Human geneticists are prohibited from or restricted in how they manipulate the germline (stem cell research).

Consequently, any genetic-based treatment of a problem in humans using rDNA technology dies with the patient and is not transmissible to their offspring.

Tissue culture is a very important technique in modern biotechnology. This entails growing plant or animal tissues and cells on artificial medium under laboratory conditions. The technique can be used to multiply plant tissue. The plants resulting from the method are genetically identical (clones) because their reproduction did not include the sexual process (allowing genetic recombination to create diversity). Clones produce uniform products and are desired in developing crop varieties where the consumer or processor desires such uniformity.

However, when clones are used in production, the downside is that, should there be a disease incidence, the entire crop can be wiped out (lack of buffering from genetic diversity). Whereas clones are routinely produced in plants, the reader may recall the social condemnation that greeted the application to animals with the development of Dolly the sheep in 1996. This significant scientific achievement was accomplished by a method called Somatic Cell Nuclear Transfer, whereby DNA from a somatic (body) cell (specifically from the mammary gland) was obtained and inserted into the hollowed-out egg cell of the sheep. Since then, clones of mammals have been produced in many species without fanfare or protestation, as the novelty has worn off.

The genetic marker technique is used widely in biological research and applications. A genetic marker is a gene or DNA sequence with a known location on a chromosome used to identify individuals. During the selection process previously discussed, the breeder literally discriminates among the hundreds and sometimes thousands of genotypes produced from the cross visually to identify the individuals with the desired genes. But how do you know the genotype you select has the desired gene? Remember, genes are expressed in an environment, which affects the phenotype. So, how does the breeder know if the genotype selected is, for example, tall because it has genes for height, or it was lucky to find a spot with high fertility in the field? A genetic marker linked to a gene of interest will ensure that the influence of the environment is eliminated. The marker is easier to assay and speeds up the breeding process. These days, instead of just a segment of the DNA on a chromosome, DNA sequencing technology allows scientists to obtain the entire sequence of DNA of the organism (genomic sequencing). You may have heard on the news that the DNA of the COVID-19 virus had been sequenced. This would enable scientists to compare viruses from various regions of the world or within the US to see whether the virus has mutated from the original infection. If the virus is mutating rapidly, developing a vaccine against it would be like trying to hit a moving target. The cost of DNA sequencing is quickly declining to make it readily accessible to researchers.

In humans, some of the significant applications of molecular markers in medicine include identifying breast cancer and muscular dystrophy using the BRCA I and BRCA2 genes (BReast CAncer genes) and DYSF, respectively.

If a genetic analysis of a woman's DNA shows the presence of one of the breast cancer genes, it indicates that the person is predisposed to the disease. Following such a diagnosis, some women opt for removing both breasts.

In this chapter, I have provided the viewer a peek into the world of scientists, from how decisions are made as to what to research and some tools and methods used in solving problems or developing products for use by society. I have also demonstrated that the government has your back, so to speak, for scientists must observe safety and ethical principles in their work, so society is not put at risk. I hope you are now better informed to participate in discussions on science and how it impacts society.

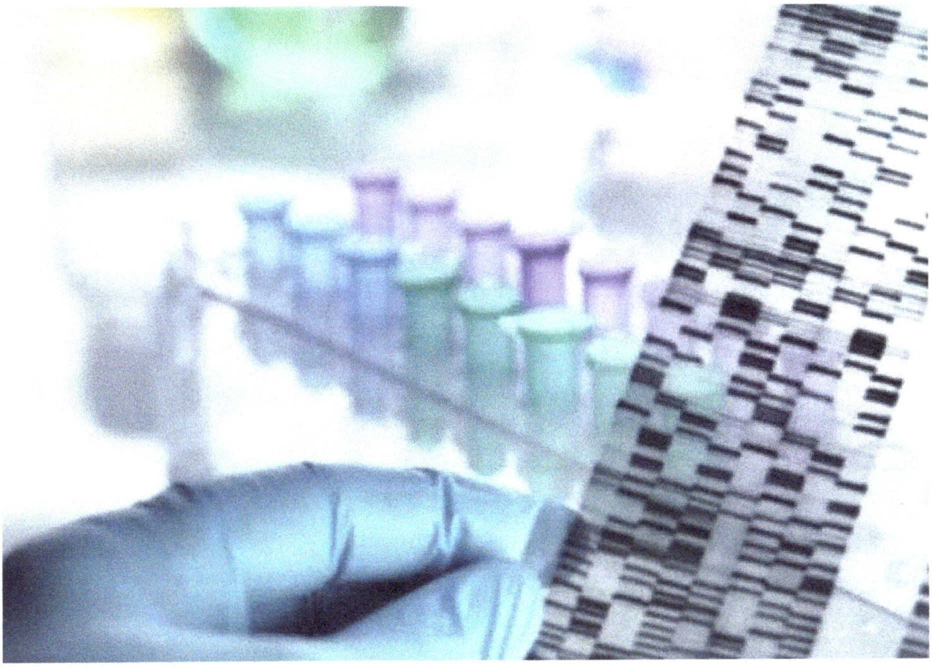

CHAPTER 7

MIND POWER AND STEM

In this chapter, I discuss a mixed bag of biological and philosophical concepts that impact how society engages STEM. The mind is critical to the practice of both faith and science. The two domains are guided by truth, a concept defined differently by each of them. Then, society weighs in with the concepts of morality and ethics to impact how STEM operates. Have you ever heard of the allegation of "playing God" that is sometimes leveled against scientists? What about society's disapproval of stem cell research? An understanding of these concepts would be helpful in properly participating in discussions on such subjects.

Some Christians have a problem with the fact that it is okay to indulge in critical thinking as we live the Christian life. Asking questions when we read the Bible is viewed as wrong and tantamount to doubting the infallibility of Scripture. But how can we probe and understand nature (in other words, do science) without asking questions? As Galileo pointed out, God gave us our minds for a purpose. We are not robots but thinking beings. In fact, the Bible says we have "free will" to make choices, which requires that we think through our decision making process. Also, it is with the mind that we serve the Lord (Romans 7:25). Christians should learn to live a full life as reasoning and intelligent beings.

But what exactly is the mind? How important is the mind to a person's existence? How important is the mind to Christian living? These are not meant to be profound questions that should keep people awake at night.

Many more questions about the mind will emerge as we proceed. Let us continue with listing some common expressions in society that reference the mind. "Mind your own business." "You must be out of your mind." "Please, keep in mind..." "You have a dirty mind." " I lost my mind.""A mind is a terrible thing to waste" (UNCF slogan)."Try to get it out of your mind." "Use your mind." "This is a case of mind over matter." "You can do anything if you put your mind to it." "Out of sight, out of mind. "These, and their variations, are some of the many ways the term "mind" is referenced in society.

The average person will point to the head as the part of the body associated with the mind. The head houses the brain and therefore is the biological organ (physical and tangible body part) associated with the mind. But the brain (tangible) and the mind (intangible) are not one and the same.

One of the definitions I found is that "The mind is the set of thinking faculties including cognitive aspects such as consciousness, imagination, perception, thinking, judgment, language, and memory, as well as noncognitive aspects such as emotion." By this definition, the mind is a pretty busy entity in our lives.

It is said that the mind is the seat of consciousness and is practically the essence of our being. It enables us to be aware (conscious) of the world around us. Without the faculty of consciousness, a person cannot be considered meaningfully alive, even though they retain the capacity to breathe (be it voluntarily or mechanically aided). Scientific research on the subject has led to various concepts and theories of unity of brain and mind. That is, neither one occurs alone, but rather each occurs, exists, and functions with the other while maintaining their fundamental differences of state (material vs. nonmaterial).

Additionally, the mind is considered a very advanced and complex information processing entity. Put another way, we humans are basically informational entities. We regularly take in and process information through our natural senses (taste, smell, sight, hearing, touch), thoughts, emotions, etc., and act upon them (think, plan and command). Some of these mental phenomena (e.g., vision, sound, and odor) are governed by physical laws and are physically predictable.

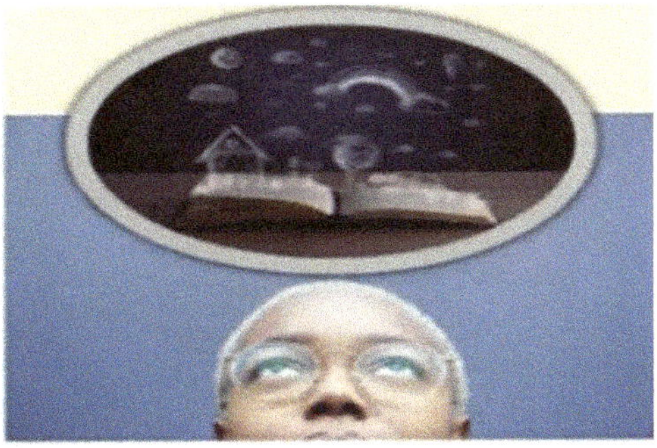

One of the well known researchers and contributors to the modern theories of the mind is the Austrian neurologist Sigmund Freud. Three levels of awareness are recognized in science as follows conscious, sub conscious, and unconscious. Proportionally, the conscious mind constitutes about 10% of the mental capacity, while the subconscious is approximately 50-60%, and the unconscious is about 30-40%. Some researchers simplify the levels into two broad ones: conscious mind (12%) and subconscious mind (88%). The conscious mind is what a person is aware of at any particular point in time. This is where logical and intellectual processes (thoughts and perceptions) occur. This level of the mind is where decision-making. planning, and communication through language are controlled. It is also associated with shortterm memory. It does not depend on accessible information to function (functions with just in time or as-you-go information).

The subconscious mind is just below the surface of awareness and depends on stored information that can be readily accessed or retrieved. This is the location of long term memory or recollection (e.g., your wife's birthday you don't want to forget that, the last movie you watched, the time it takes to drive to work). The subconscious mind controls mainly the physical functions of the body. It includes functions like breathing and digestion, as well as emotions, feelings, gut instinct or intuition, spiritual connection, habits, addictions, fears, and relationship patterns. The unconscious mind is buried deep below our conscious awareness. While we don't actively access this information because we are not aware it exists, it nonetheless can significantly affect our behavior (e.g., a very bad experience you have forgotten about, an intense dislike for someone like for a parent). It's where we store our experiences, beliefs, and every image we have ever seen, among other inputs and stimuli.

As previously stated, the mind is believed to be a set of thinking faculties. Traditionally, people are expected to make rational, reasoned, and logical decisions. However, humans experience and exercise a form of intelligence that is beyond the rational mind. We have the ability to understand things instantly, unaided by any amount of conscious reasoning.

Called intuition, it has been described as the ability for direct knowing or an inner knowing that comes without one thinking it through (i.e., reason is interrupted). Society has a number of expressions for this human capacity, including instinct, sixth sense, extrasensory perception, and second sight. It is common to hear people say things like,"I had a hunch," "I had a gut feeling," "I heard an inner voice," "I had a feeling," "My heart told me," "I had an aha! moment," "and "I knew instantly." Have you ever been in a situation and felt immediately that something was not right? Or, have you heard the phone ring and knew who was on the line before they introduced themselves? These contextual and relational experiences are examples of intuition. One uncommon way of intuiting information and knowledge is practiced by psychics and those who dabble in clairvoyance (the claimed ability to gain information about an object, person, location, or physical event through extrasensory perception).

Intuition is an innate capacity (natural process) present in all humans. Just like logic is not always right, intuition can be faulty at times, the latter resulting when our thoughts and emotions taint the experience. Consequently, it is believed that intuition can be "perfected" or improved (developed) and made "purer or truer." Many cultures see intuition as fundamental to religious and spiritual experience. Many people believe in a power or force beyond themselves, often called the divine. Christians and some other religions recognize the spiritual being called God, whom we engage through prayer and intuitive experiences like worship. The Bible teaches that Christians should heed "the still small voice" in themselves. This is the voice of God that can be heard loud and clear when the believer is calm and still, having tuned out distractions (e.g., wishful thinking and projections), which cause them to substitute their ideas, fears, and what they hope for with true intuition. One has to practice God's presence, learn and become sensitive to the prompting of the Holy Spirit, and obey instantly. Before I draw the curtain on this topic, I would be remiss if I did not mention the phrase "counter-intuitive."When an outcome or expectation goes against what is logical or common sense, it is said to be counterintuitive.

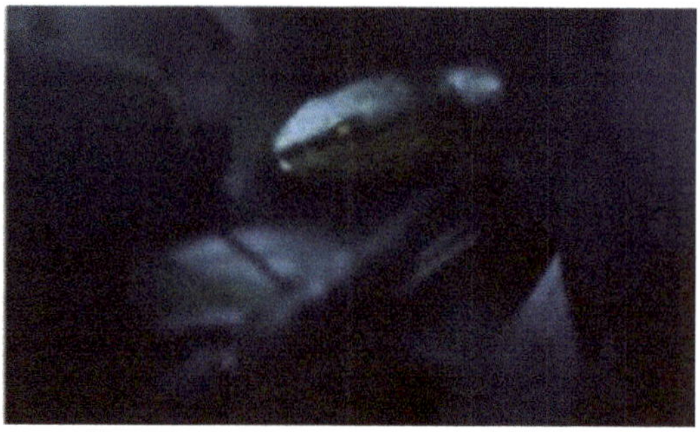

On the religious front, the mind is central to Christian living. The mind is described as a battle field on which good and evil forces do combat for control over it, and thus control over one's life. Our enemy, the devil, desires control of our minds and plays mind games to accomplish his purpose. His playbook in includes sowing seeds of doubt about what we believe, planting negative thoughts of all kinds, and instilling fear in us. According to Scripture, these deceptive tactics started in the Garden of Eden, with the temptation of Adam and Eve by Satan. I once heard a preacher say, "As the mind goes, so goes the man."The Bible has teachings on overcoming the enemy on the battlefields of our minds. We are told to love the Lord our God with our hearts, souls, and minds. As previously stated, it is with our minds that we serve God, so once the devil has control of our minds, we serve the law of sin. God has promised to keep us in perfect peace if we keep our minds stayed on and trust Him. God does not play mind games with his children, for all of His promises have been fulfilled in Christ. The Bible has stated clearly that all things are possible to the believer through Christ, and if we don't doubt when we pray to God, even mountains can be moved.

The mind processes information that we take in through our senses (sight, taste, sound, touch, smell) and our thoughts and emotions, etc. That is why a popular Sunday school song cautions us to be careful what we see and hear. God gives us the spirit of a sound mind and not fear.

On the battlefield of the mind, we do not fight in our own strength or use carnal weapons but spiritual ones.

The prescribed spiritual armor is detailed in Ephesians 6:10-18. It is interesting to note that the head, the residence of the brain and our mind, where thought and knowledge reside, is to be protected with the helmet of salvation. If we doubt our salvation, we doubt our relationship with God, and we have no power over sin and Satan. The devil is a liar who plays mind games rooted in deception. The Bible admonishes us to constantly renew our minds by washing them with God's word (Romans 12:2). I call this spiritual (Biblical) mind therapy.It is more commonly called meditating on God's word. Christians have the mind of Christ and are imbued with divine power to demolish arguments against the knowledge of God successfully. When negative thoughts come into the mind, we must act immediately to arrest and bring them under the authority of Christ.

Thinking intuitively and analytically both have their place in the life of a Christian. Yes, Christians are thinking beings. At the very least, we have to exercise common sense in our decisions in life. But a powerful strategy Satan uses is to question the validity of our belief or faith by encouraging us to be analytical and intellectual about it. We are tempted to apply reasoning and thinking to faith. Faith does not always make sense; human sense, that is. Some research suggests that people who rely more on intuition in decision making are more likely to believe in God.

On the contrary, thinking analytically reduces people's tendency to believe in God. The methods of faith and science are diametrically opposed. STEM is advanced and applied primarily through analytical thinking, but intuition has a role in scientific and technological breakthroughs. There are examples in society where people have been dismissive of faith or religion early in life because they were not always supportable with logic and analysis. But later in life, they saw the light, so to speak, and became believers. Of course, the reverse also occurs for various reasons.

Another brain related issue that society sometimes confronts is the state of being brain dead. Medically speaking, a person who is declared brain dead is legally dead with no chance of recovery. The body at this stage is unable to survive without artificial life support. This is different from being in a coma, which is being unconscious but still alive. An active debate in society is about who has the right to pull the plug on a life support system, thereby definitively ending life.

Some have likened the subconscious to the cruise control on a car or autopilot on a plane. These devices are preprogrammed to perform in predetermined ways until the instructions are overridden with new ones. Programming of the subconscious occurs through influences in nurture (environment in which we develop), people with whom we interact, TV programs we watch, religious experiences, etc., such that solid foundations are laid down by the time a kid reaches age eight or so, according to experts. In early life, the brain just soaks in whatever information it receives, including, unfortunately, negative ones (e.g., you will never amount to anything, you are stupid, your race is inferior, women cannot perform as well as men, etc.). The pieces of information, the good, the bad, and the ugly, are stored away for future reference in later life.

Whatever we plant in our subconscious mind and nourish with repetition and emotion will one day become a reality (Earl Nightingale). Because the information is stored below the level of our conscious awareness, we become aware of its existence when it manifests by adversely impacting our progress in life. Fortunately, we have an opportunity to reprogram our minds (change the brain). In other words, we can change our minds and thereby transform our lives. You have to replace the negative ideas with positive ones (I am good enough, beautiful, smart, and can do anything I put my mind to). As noted earlier, the Bible admonishes us to reprogram our minds daily by washing them with the word of God. This will keep the old sinful nature from resurfacing. Contrary to the popular saying in society, "You cannot teach an old dog new tricks," it is never too late to change your mind. It might be more challenging to change people we believe are "set in their ways."

You may have heard the phrase "mind over matter." This is usually invoked when individuals use their minds (will power) to control a physical condition or problem and keep going even when exhausted. There is significant evidence to show that the mind can heal the body to a degree. Positive thinking initiates biochemical changes that positively impact physical health, including reducing pain to improving the immune system. One of the most dramatic demonstrations of this human experience occurred in the life of one of the most brilliant minds of recent times, Cambridge University's Stephen Hawking. A theoretical physicist and cosmologist, Dr. Hawking was stricken with a debilitating form of ALS, which left him wheelchair bound for most of his life. Despite these incredible odds, he made some of the most outstanding discoveries toward understanding our universe, communicating only through a custom voice synthesizer while on a ventilator. He is often compared with high-caliber scientists like Albert Einstein and Isaac Newton. He succumbed to the disease in 2018. His family donated his personal ventilator to the hospital that took care of him to be used to aid COVID-I9 patients. In one of his last interviews on Star Talk, hosted by the famed astrophysicist Dr. Neil de Grasse Tyson, Stephen Hawking's closing remarks to the world were "Never give up."

One of my favorite past times or entertainment choices is illusion, commonly called magic. I love music, and I love to play and compose songs. But my next love is magic, not that I can create major illusions at this time. Magicians create tricks or illusions by taking advantage of how humans perceive external stimuli (taste, touch, sight, etc., and our minds) and process information. Their trade has ancient origins. Modern day scientists (neuroscientists) have studied how magicians accomplish their seemingly impossible illusions with such cheeky ease. Illusions separate perception from reality. Illusionists use three basic techniques optical, mechanical and psychological to take advantage of how our nervous system works. They masterfully manipulate the audience's attention, which they misdirect overtly (direct the audience to look away from the location where the trick is actually occurring), or covertly (by subtler manipulation). Some politicians and religious cult leaders, use similar tactics to manipulate their followers. Such leaders are often charismatic, and sometimes even ruthless. Repeating falsehood over and over for a period, eventually becomes accepted as truth. This tactic was on full display during the 2020 presidential elections. The social media world provided a fertile ground for conspiracy theories and dis/ misinformation to be propagated.

The role of the mind in our lives as human beings cannot be overemphasized. It is heartening to know that we can reprogram or rewire the mind for a renewed purpose. Our brains are wired for thought; we are thinking beings.

We take information into our minds and transform it into knowledge we act upon. Without knowledge, we shall perish, both in our spiritual and social lives, for we shall make uninformed decisions. In addition to the Bible, I encourage Christians to pursue science literacy, so we can probe knowledge and information to know or understand objective truth in our world. Some of the existential questions that religion seeks to answer are: Where do we come from? What are we? Where are we going? There is a lot of living between the beginning and the end. The Bible says we shall give an account of how we live.

One of the problematic issues in society that implicates the mind is bias, the prejudice in favor of or against a thing, person, or group compared with another, and usually conducted unfairly. Prejudice is defined as a preconceived opinion not based on reason or actual experience. Bias can be explicit (in your face), whereby the individual is aware and means to conduct themselves in that manner (e.g., I don't like Black people). It can also be implicit, in which case the individual is not conscious of their behavior (e.g., because you hate Black people, you always vote against a Black candidate on a ballot). Implicit bias operates subconsciously and results from attitudes and stereotypes an individual harbors that affect one's understanding, actions, and decisions. It is activated involuntarily and without a person being aware or able to control it intentionally. The danger with implicit bias is that we all have it!

I read an article by Graham Lawton of the New Scientist, entitled "The grand delusion: Why nothing is as it seems," Which discussed how the mind plays games with us. As I previously stated, we engage nature and the world in general through our five senses. Graham states that our sensory perception, especially visual, is but a figment of our imagination. According to psychologists, what we experience is largely the product of what is inside our heads. What we experience through the eyes is a tiny fraction of the total visual information available at the time. Ironically, eyewitness testimony is widely accepted and, in fact, encouraged by the courts of law when scientifically speaking, it is one of the worst and most unreliable pieces of evidence in a case. I am not referring to false eyewitness testimonies to start with and given out of prejudice. Some eyewitnesses come forward sincerely believing that they saw what they think they saw. Many inmates are languishing in jail or may have been executed based on eyewitness testimony. Ironically, it took many years for the courts to accept DNA evidence as reliable.

On the matter of prejudice, researchers believe that we all look at the world through a veil of prejudice and self-serving hypocrisies. We have a conviction that no one but us perceives the world as it really is and that anyone who sees it differently is biased (called naive realism). On the other hand, if you fail to accept this statement I just made, you suffer from yet another kind of illusion that professionals call bias blind spot (you acknowledge that such biases exist but only in other people). Implicit biases are laid down in childhood and early adulthood but lie low in our subconscious. Because of where biases reside in the brain, they are not consciously accessible for introspection even when we look inwards to question our judgments and beliefs. This leaves us to falsely conclude that our beliefs are adequately reasoned and correct.

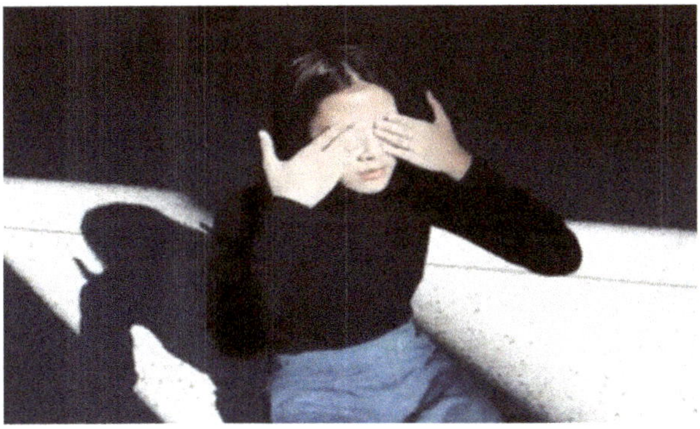

As humans, we harbor many innocuous biases that protect our egos from reality. We tend to take credit for success but deny responsibility for failure. The more harmful biases that we continue to battle in our American society include racism and sexism. When these touchy subjects come up, it is common for many to deny they have such biases while their actions betray them. Understandably, personal opinions could suffer from bias. Still, unfortunately, truth or fact can suffer the same fate when we try to interpret the world to fit our existing beliefs, something some Christians and politicians are adept at doing. They seek information to bolster their position (self serving) and dismiss anything that does not fit their narrative as a hoax (called confirmation bias). When this happens in politics, opposing parties dig their heels in and refuse to budge or entertain any compromise. The most concerning thing about bias is that you could be swimming in it and not be aware you are doing it. Sometimes, when people are confronted with new information about their bias, it causes them to dig deeper and harden their position on the subject.

Our biases strongly influence the way we accept science in society. Politically, the Conservatives and Liberals have favorite positions on major social issues associated with science (e.g., climate change and evolution), which they fiercely defend. These biases are sometimes rooted in fear or lies, and misinformation.

(For further reading: Steve Pinker (1997) *How the mind works*. W.W Norton and Company).

CHAPTER 8

HOW FAITH WORKS

To distract from the goal of this book, I have selected Christianity, the tradition with which I am most familiar and the most popular globally, to guide all discussions going forward. Before we proceed, I also need to clarify some terminologies, specifically faith, belief, and religion, which are often used interchangeably in society. Let me use basic dictionary definitions for these terms. Faith is defined as complete trust and confidence in someone or something, and is often associated with religion (no proof required). Belief is generally seen as a firmly held opinion, conviction, or confidence in a person or thing.

On the other hand, religion is a specific belief system with codes of conduct, ethics, and philosophy. We may sum these up by saying faith is a religious belief. If unsatisfactory to you, please study the matter further to understand faith better, for it is absolutely critical to being a Christian.

I have heard many preachers emphasize that Christianity is not a religion but a personal relationship with Jesus Christ. This is true, for it is first an individual experience before it becomes a group experience. You are saved as an individual and go to heaven as an individual and not by affiliation with a group (church, family, etc.). There is no "religious herd immunity" protection from going to hell; you must have your own vaccination, if you will. As indicated in the third chapter, religion is human made. No wonder there are thousands (over 30,000) of Christian denominations in the world, each with its defining tradition or culture. Unfortunately, these traditions are a source of division among denominations and often overshadow the true essence of Christianity. These days, it is not enough to say you are a Christian when you meet another believer.

The questions from the inquiring mind that most often will follow include: "What denomination do you belong to?" "What church do you attend?" "Who is the pastor of the church?" "Is it a White church or a Black church?" Based on the responses to these questions, opinions will be formed about your "genuineness" as a Christian, for each denomination (and church or pastor) is preceded by a reputation in society, for better or for worse. Some are judged as cult like, rightly or wrongly, while others are viewed as political or social groups masquerading as religious entities. The Bible cautions that our traditions can lead us astray as Christians and nullify the word of God (Mark 7:13). That is a pretty scary caution that we must heed.

We are warned against adding to or taking away from Biblical truth or scripture. In short, do not make Christianity more or less than it is. Scripture admonishes each one to work out their salvation, study the word of God and rightly divide or interpret it. Of course, we need pastors and leaders who do their due diligence in studying God's word so they can teach it unadulterated with human wisdom and traditions. In light of all that is going on with Christianity in America today, especially in the last decade or so, I prefer to say simply that I am a "follower of Christ." Denominations and other divisions mean little to me. All that matters to me is that I may know and follow Him.

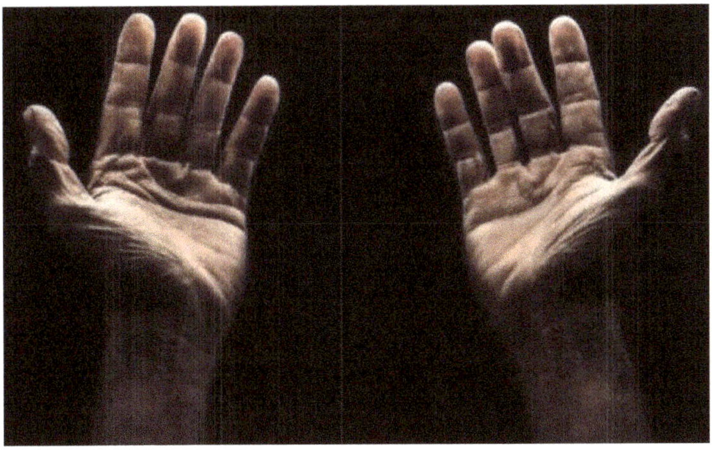

The Holy Bible guides Christianity as its sole authoritative guide to spiritual truth. God is truth, and the Bible is believed to be His inerrant word. Let me take a minute to retrace the Christian faith journey.God created the universe with a purpose for humans to enjoy a close fellowship with Him. The purpose of creation took a different turn when Adam and Eve exercised their divinely ordained right to choose and fell victim to the deception of Satan (the devil), and with that, evil (sin) entered the world. This relationship was severed, resulting in humans being banished from the Garden of Eden to fend for themselves. Subsequently, humans had to contend with two voices, God's and Satan's, the voice of truth vs. the voice of deception (lies). With sin now part of their experience, the offspring of humans and the society in which they lived were inclined to evil. The people progressively moved away from God, wallowed in all imaginable evil, and were no longer responsive to Him. Finally, God decided to destroy the earth with an epic flood after a 120- year grace period, sparing only Noah and his family for being the ones who inclined their hearts to God.

When the flood receded, this family assumed the role of Adam and Eve to repopulate the earth.The new groups of humanity continued in the wicked ways of the first, eventually challenging God in the epic rebellion at Babel. Following this standoff, God identified Israel, who did not rebel against Him, as His people through whom His plans for reconciling humanity to himself would proceed. The Bible declares, "But God so loved the world, that He longed to reconcile fallen humans to himself (John 3:16)." This is one of the most comforting scriptures in the Bible. Eventually, the way back to God is through faith and belief in Jesus Christ, personally accepting Him as Lord and savior.

Yes, all humans are creations of God.Yes, we are all His children, except we are born estranged from and out of fellowship with Him. Through rebirth (born again) in Christ, we are given the power to become children of God with benefits and opportunities that are unavailable to those who remain estranged from Him.

I will keep this discussion of faith and religion very simple and not interject any unnecessary theological concepts. Just as with the discussion on science, the reader may want to pursue some of the theological concepts I broach on their own for a deeper understanding as desired. I will start with some key doctrines in Christianity. The God of Christianity is a Triune God (Trinity), comprising co-eternal and consubstantial persons:the Father, the Son, and the Holy Spirit. Though distinct, they constitute one substance, essence, and nature. Another doctrinal notion is that God the creator is immanent and transcendent with regard to His creation. The immanence attribute means that He moves or operates within His creation (He is part of it). Transcendence refers to His nature and power independent of the material universe and beyond all known physical laws (i.e., His creation does not constrain him). A third essential doctrine is divine predestination. God predetermined every word, thought, and action we would ever take before creation (i.e., all events have been willed or preordained by God; He is omniscient). Yet, God has granted humans the free will to choose as they desire. This paradox makes God's omniscience seem incompatible with human free will.

One explanation I have heard is that the simultaneous presence of omni science and free will in the universe prevents humans from placing all responsibility on God, and thereby becoming legalistic: or assuming that all depends on us, leading us to do as we please. Predestination not with standing, God holds us accountable for our decisions and actions. In essence, our choices include our personal choice, God's influence and the influence of Satan. These are deep theological concepts that you are encouraged to pursue further on your own.

Now, let me turn my attention to some simpler doctrines in Christianity. When we talk about the Bible being the inerrant word of God, we mean that its truths do not change with time. The principles and concepts therein transcend time. As noted in the introductory chapter, the contexts in which these truths were established or given are mostly ancient agrarian cultures, which we now apply in our ever-changing modern high-tech cultures. We should recognize that faith or religion are supposed to change society and NOT the reverse. The way we engage a constantly evolving society through religion (a human construct) may vary, BUT THE CORE MESSAGE (principles and concepts) does not change. The Bible is not like a textbook that some of us write, which requires periodic review and revision because some information may be outdated. This places the onus on us as Christians to review and revise our METHODS of engaging people with the Gospel so it remains relevant in our EVER-CHANGING society. It is crucial that we understand and properly embrace science (STEM), the driver of the evolution of our modern society. It makes no sense whatsoever to be wedded to archaic traditions or the new ones we create, which are unhelpful to our task of reaching out to the world.

Another central theological doctrine in Christianity is the concept of agreements or covenants. God engaged His people Israel under a Covenant. When Jesus came on the scene, He inaugurated The New Covenant and declared it superior to the Old Covenant, the latter being administered under hundreds of laws. Previously, when a law was broken, it was atoned for by sacrificing animals. Under the New Covenant, Christ died once and atoned for all the sins of humanity with His own blood (the Sacrificial Lamb). This means, in Christ, our past sins, present sins, and future sins are all paid for in full. We live under grace and not law. If we falter, we can repent and ask for forgiveness of our sins.

Christianity stands separate from all other world religions in several respects. First, Christianity declares that Jesus is the only way back to God. Any other way leads to eternal damnation in hell. If that is not enough pressure on us, Christianity is a proselytizing faith. We are commanded to go into all the world and make disciples of Christ (Mark 16:15). According to this doctrine, everyone is wrong to start with, so when you meet a person of no faith or another faith, you have your work cut out for you, trying to persuade them to accept YOUR faith or position. This is a challenging position to be in. Christians must therefore engage people in love and with grace, patience, respect, kindness, and humility. Insulting, shaming, bashing them on the head with Scripture, and other religious practices are unbiblical and counter productive. It is critical to note that the Holy Spirit convicts people of sin and leads them into repentance and salvation.

As fishers of men, our role is to cast our net wide into the waters, drag in the haul, and place it at the feet of Jesus, who will clean the catch. It is impossible to live the Christian life on your own. I believe a Christian can say," I have been crucified with Christ and risen to a new life (born again) in Him. It is no longer I who live, but Christ lives in me. So I live in this earthly body by trusting in the Son of God, who loved me and gave Himself for me" (Galatians 2:20). Christ summed up the New Covenant this way: "Love the Lord your God with all your heart, and with all your soul, and with all your mind. Love your neighbor as yourself" (Matthew 22:37-39). Your neighbor is anyone other than yourself. LOVE is the mother of all principles or doctrines in Christianity. Love all people unconditionally. "If we say we love God but hate others, we are liars, for we cannot love God, whom we have not seen, if we do not love others, whom we have seen" (1John 4:20).

Yet another Christian doctrine is transformation. Being born again is the easy part, but being TRANSFORMED into the image of Christ is a process that takes time and varies for each person. You can be a Christian and still struggle with certain old sinful habits. But whatever sin you came into God's family with is forgiven and forgotten. Do not beat yourself up over the past or allow anyone to do the same to you. The devil is the accuser of the brethren (Revelation 12:10). Be careful how you judge others. There should be no place for vigilantism in the church, whereby we wrongly select a few sins and label them as "Cardinal sins" or perform a "litmus test" based on human traditions and harass people. Instead, we should allow the Holy Spirit to do His work of changing hearts and transforming people. We must deal with people with grace, but by the grace of God, it could be you wallowing in the filth of the sin you are pointing to. Some of the "vilest offenders" in Scripture were mightily used by God for His purposes. He did not condone their sins but dealt with them graciously. God looks upon people's hearts and judges their intentions rather than actions. This speaks to the character of God being loving and merciful. When the adulterous woman was about to be lynched by religious zealots, Jesus came to her defense instantly. He said to them if you have no sin in your life, go ahead and cast the first stone (John 8:7). We have to love people into the family of believers or the church, not through hate and fear.

There will always be sin in the church because God's transformation work is ongoing. It is fair to say that any sin in the world (i.e., outside the church) will likely be present in the church in some shape or form because people are still being "worked on" by the Holy Spirit. But let me hasten to add that the church is not a safe haven for people to continue to practice sin. Instead, it is a place where Christians learn to grow and draw closer to God and, in the process, shed their sinful habits. In Christ, we are a new creation; old things are passed away and replaced by the new. You may sometimes hear unbelievers accusing Christians of being hypocrites when they point fingers at sin outside the church while it goes on inside it. The more reason why we should be careful not to be judgmental and thereby keep unbelievers at bay. Do not look past the plank in your eye and try to pick the speck in another's eye (Matthew 7:3-5). Some will not let you into their church or fellowship unless you pass their "eligibility test"

(tradition). This is like saying to a patient, go and heal yourself, then come to our hospital for evaluation.

While we are being transformed into the nature of Christ, we must go about the business of a Christian. Christ came to the earth to do the will of His Father in heaven, which is to restore sinners to God so that they may have eternal life with Him (Matthew 7:21). This is what I call the "big picture of Christianity" (seeing the "forest"). We must be careful not to let "trees" (e.g., our traditions) get in the way of this rescue mission. What else is dear to God's heart? The Bible teaches that righteousness and justice are the foundation of the throne of God, while truth and mercy go before Him (Psalm 89:14). This verse speaks to God's character. Some Bible translations indicate that "steadfast love and faithfulness" go before Him. So, God cares about righteousness, justice, truth, mercy, love, and faithfulness.

Some explain that justice is making things right that are unfair in a nation or society, while righteousness is doing right by the people, especially the vulnerable (Steve Holloway). A Bible verse made popular in modern times by Dr. Martin Luther King, Jr. is: But let justice roll down as waters and righteousness as a mighty stream (Amos 5:24). What God wants and expects of Christians is for us to practice justice and right living, and not just go through religious motions or traditions for their own sake (what some call "doing church"). Who are the vulnerable in society, by the way? These include the widows, orphans, immigrants, and the poor (the so called "quartet of the vulnerable"). When we turn a blind eye to the plight of these people, we display a lack of Love and mercy, and God sees it as injustice. In our modern society, when people say that they "want to see justice done," it is often about crime and punishment.

Even in this situation, what God calls for is equity and fairness. Do not pervert justice; do not show partiality to the poor or favoritism to the great, but judge your neighbor fairly (Leviticus 19: 15). Defend the cause of the weak and fatherless; maintain the rights of the poor and oppressed. Rescue the weak and needy; deliver them from the hand of the wicked (Psalm 82:3-4). He who oppresses the poor shows contempt for their Maker, but whoever is kind to the needy honors God (Proverbs 14:31). The only person who belongs on any law court, supreme or not, is the one who will administer justice without partiality. So, this political tactic of "packing the courts" or appointing justices of a political stripe (liberal or conservative) may be good politics. Still, it is not Christian, and it is not Biblical.

Just like justice, when righteousness is mentioned in our modern society, the tendency is to associate it with personal morality (honesty, sexual purity, reverence, etc.). This has given rise to what I call "sin policing" in the church, whereby some even go to the extent of guilt shaming others who are struggling with certain sins. According to the Old Testament, where many of these scriptures on justice and righteousness originate, the latter is more about how we live (right relationship) with others in society treating all people with fairness, generosity, and equity. "It is 'primary justice,' a behavior that if it were prevalent in the world, would render rectifying justice unnecessary, because everyone would be living in right relationship to everyone else" (Tim Keller). Righteousness and justice are joined at the hip, so to speak, in society and not by accident. In modern society, they are often interpreted to mean "social justice," an expression that is highly politically publicized. It distracts from the truth it contains and challenges politicians and society in general to act. The Lord loves righteousness and justice: the earth is full of his unfailing love (Psalm 33:5). "Let not the wise man boast of his wisdom or the strong man boast of His strength or the rich man boast of his riches, but let who boasts boast about this: that he understands and knows me, that I am the Lord, who exercises kindness, justice, and righteousness on Earth, for in these I delight, "declares the Lord (Jeremiah 9:23-24). So, here you have it. God delights in justice and righteousness in society, and so should you. You can't explain it away.

Another key doctrine in Christianity is faith. It appears to be one of the least understood concepts and most misapplied by the faithful. Ask the average Christian what faith is, and you'll most likely get the popular quote, ".... the assurance of things hoped for, the evidence of things not seen (Hebrews 11:l)." Okay, but what does it mean? What really is faith? What does it mean to walk by faith and not by sight? For some, having faith in God is dispensing with the role of the mind - i.e., no thinking or reasoning through things allowed. Just accept things at their face value and do not ask questions. For others, faith is like a magic wand that you may use to conjure something into existence. Also, faith operates only in the supernatural realm and the miraculous. Taking medicine is anti - faith; prayer is all you need. Just name it and claim it, by faith, and it shall be yours. For some, faith is a perspective.

We get our righteousness or right standing with God as a gift from God Himself. Our righteousness is like filthy rags before Him. An unbeliever does not benefit from the free gift of righteousness, so we must deal with them with grace and mercy, for by the grace of God, we could be in worse shape. If an unbeliever behaves contrary to church doctrine or does what you consider sinful, it is normal for them. There is no need to pounce on them; they don't even have the benefit of the Holy Spirit to assist them in living right, so don't judge them by your standards. Instead, love them unconditionally, and pray for them to come to know Jesus. As I said previously, religion is supposed to change society and not the reverse. That is why it is absolutely critical for Christians to know the real truth of the Bible first and understand the nature of society so we can influence it properly and effectively. Society is evolving, but the truth of the Bible remains immutable, at least the essence of it, as said before.

We must adjust how we engage society so we do not drive away those who need the Message by insisting on our traditions. Yes, Sunday best was the traditional attire for a church in the past and is still popular in many churches today. But these days, T-shirts and ripped pants are standard issue for the youth and young at heart. God looks on the heart and not the outward appearance. So, don't deny anyone the opportunity to hear God's word by how they are dressed. Without hearing God's word, there can be no faith.

As humans, we grow in our faith as we daily are being changed into the image of Christ. So, it is common for people, even preachers, to grow out of or alter old positions on some theological matters as they have new revelations on them. We cannot change the message of the Bible to accommodate a changing society, but we must change how we engage society with the message of Christ.

We talk about America as having been founded by Christians, but we now lament that the country is going in the wrong direction, spiritually and morally. I question if our witness has been effective over the hundreds of years of its founding.

Have our words matched our actions? Do we practice what we preach? More importantly, do we know the truth of the Bible, so we can practice it correctly as it should be, rather than our tradition tainted truth or outright untruth?

According to a 2020 Pew Research Center report, 65% of Americans profess Christianity as their religion or faith (down from 85% in 1990). This may be interpreted in various ways. America is a nation of immigrants from all faiths and all comers of the earth. As more non Christian immigrants come in, the proportion of Christians can go down. This should be good news for Christians in fulfilling the Great Commission of Christ to go into all the world and preach the gospel (Mark 16:15). Now, the world is coming to us in America! But a more critical factor for this state of affairs, in my opinion, is Christian conduct (or shall I say misconduct?). Are we taking advantage of our situation and reaching out to the unsaved amongst us? Or are we blaming others for the reality in our nation? Do our actions in society reflect true Christianity worthy of emulation, or do they contradict WWJD (What Would Jesus Do?). What does Christianity mean today - affiliation with a political party, a tradition, hate, or love?

It is absolutely crucial to read and study the Bible in context. Because of the times in which Jesus lived on Earth, His examples were based on things and customs of those days. Agricultural examples abound in the Bible because that was the prevalent occupation. Just because it is written in the Bible does not mean it applies to us or should be emulated. Just because we are Christians does not mean life will be rosy and smooth because God is on our side.

On the contrary, life can get bumpy and even bumpier because we are Christians. The Devil, our archenemy, is constantly looking for opportunities to cause us to doubt our faith in Christ and disobey Him. The way we survive the barrage of demonic attacks is to put on our spiritual armor, not carnal ones, for the battle is spiritual, and it belongs to the Lord (Ephesians 6:11).

If there is one message the COVID-19 pandemic drove home clearly to Christians, it is that the church is not brick and mortar, something we should have already known. People (Christians) are the Church. In fact, the Bible declares that we are a temple inhabited by the Holy Ghost (1Corinthians 3:16). We can worship God 24/7, anywhere. The one two hour Sunday church meeting may satisfy the Biblical admonishing for us to fellowship with other believers. But Monday through Saturday is when the rubber meets the road in our lives, when we show to our brethren and the world that, yes indeed, we are followers of Christ at home, at work, at play, in government, and society in general. Clothes don't make the man (or woman) as they say; the man (or woman) is who is in the clothes. Faith and love are meaningful, actionable terms. I saw a humorous cartoon a long time ago showing a mouse pointing to a mousetrap in the backroom of a church, with the caption, "And they say they are Christians". Reading the Bible at the workplace when you should be working is a poor testimony to your colleagues. After professing to be a Christian, the quality of our work should be exemplary.

A popular quote Christians often refer to when they feel society is "going to hell in a handbasket" is, "If my people who are called by my name will humble themselves, and pray and seek my face, and turn from their wicked ways, then I will hear from heaven, and will forgive their sin and heal their land" (2 Chronicles 7:14). Then, invariably, we list the usual suspects we believe to be the culprits - prayer was taken out of school, abortion rights, homosexual rights, evolution, etc. - as the critical Christian issues of concern in society for which we must repent. The problem with this attitude is that, more often than not, those who are most vocal about these selected and implicit issues in scripture are nowhere to be found or heard when the general and weightier matters, which Jesus explicitly spoke about - injustice, poverty, discrimination, inequality, hatred, etc. - are prevalent. An appropriate and powerful reflection on this matter is a quote: "We will have to repent in this generation not merely for the hateful words and actions of the bad people but for the appalling silence of the good people."(Martin Luther King, Jr.). A vertical relationship (with God) is not enough; we must have a horizontal relationship (with people) to be true followers of Christ and not just noisy gongs.

If you know what is good to do but do not do it, the Bible says it is a sin for you. It is not enough to have a bumper sticker or wear clothes and jewelry with Christian slogans to announce your faith. It is what is in your heart and your actions that will let people know whether or not you are a true follower of Christ or a pretender. Actions, they say, speak louder than words. True Christianity is not knowing about Christ, but knowing Him and following Him. When dealing with others, it is not a matter of whether it is lawful but is it expedient (1 Corinthians 6:12; 10:23)? It might be your right in our society to do or have something, but if what you do would cause your brother to sin, then refrain from such action for the sake of your weaker brother or sister.

For Christianity to be authentic, it must first be rooted in the Bible and the principles and concepts, and consistently applied. Just because it is in the Bible does not mean it applies to us. We must read the book in context to correctly apply the message therein. More than one verse or passage can bring different perspectives to the same truth. We must be careful NOT to add to or remove Biblical truth. Our traditions can tempt us to replace Biblical truth with our political positions or religious and personal preferences.

Before I close this chapter, let me broach a doctrine that is often misunderstood and misapplied free will or choice. I touched on it briefly in the introduction. God ordained free will. We live by making choices on a daily basis. Individuals may choose their course of action on issues. Choice has consequences.

So, what has faith or religion got to do with STEM? EVERYTHING! In Chapter Two, I demonstrated how pervasive STEM is in modern society. Its influence will grow progressively stronger with time as the world becomes progressively more technologically driven. We understand nature through the practice of science. Science drives technology and engineering, and eventually, the economy. We must thank God for STEM. How science is pursued and its outcomes deployed in society have ethical, moral, and hence faith implications. We must first be firmly grounded in faith to engage STEM and society in general correctly. In the next chapter, I shall discuss faith versus science and how they differ as well as work together. Until then, keep walking by faith and sight, for it is the will of God so to do.

CHAPTER 9

SCIENCE vs FAITH

Faith is intensely personal to individuals, who often fiercely defend it against encroachment from any source. I demonstrated in Chapter Three that modern society is intricately linked with STEM. Whereas Christians enjoy the benefits of science like everyone else, unfortunately, it sometimes steps on the toes of faith and religion. So, can the two domains coexist in society? If so, how can this be? Let us find out.

Science has a love hate relationship with religion and society in general. People love how science has helped dramatically transform modern society, improving the quality and quantity of life. I dare state that many people see technology and engineering and don't realize how science made it all possible. As discussed earlier in the book, the influence of science in society is inescapable, manifesting in everything around us, including the food we eat, the clothes we wear, the rooves over our heads, the places we worship, the places we work, modes of communication and transportation, and our health and leisure. However, there are certain aspects of social life in which some consider science to be an unwelcome intrusion. I believe that part of the apprehension about science in this respect stems from ignorance of what the field is all about and the improper understanding and interpretation of Scripture.

Science and religion have one thing in common - nature. Scientists try to understand how nature works and, in doing so, find out how to nudge it to the advantage of humans. Religious people are generally content with beholding the wonderment of nature (creation). Scientists want to probe nature. I mentioned in Chapter One that God has given us two books the Bible and the Book of Nature (Hugh Ross).

God authored both books. Science is all about studying nature. In fact, God wants us to study and query nature. There is a fundamental difference between how religion and science engage the world. Faith (religion) says, believe first, and then you shall see; science says, show me first, and then I shall believe. Another way of saying this is that believers walk by faith while scientists walk by sight. Science, according to practitioners, operates a culture of doubt (focus on discovering untrue things by falsification), while religion operates a culture of certainty. There are scientists like me who are believers. We have no choice but to walk by faith and by sight. So, is there a contradiction in this matter for us? Certainly not. The Bible expects Christians to walk by sight as well, for faith without works is useless. Scientists cannot do their work based on faith. If I am trying to balance a chemical equation but get stuck, I can't brush it aside and proceed with the reaction.

I believe God made me a scientist. So, I can exercise faith through prayer for His help to accomplish my task. I said earlier that we go up to space by science, but we go up to heaven by faith. It is futile and unwise to pitch faith against science. It is wrong to try to use science to prove or disprove the existence of God. As previously stated, it is impossible to prove that something does not exist. In my opinion, if you can prove God exists, maybe He does not, for the Bible says we come to or engage God through faith and not sight. But I appreciate modern scientific findings that indicate that the universe had a beginning and, therefore, by extension, a beginner, whom Christians are happy to claim to be God. There are some people for whom such arguments are powerful for persuading them to have faith in Christ and, therefore, are a good thing. Scientists cannot accept miracles in their practice. They must have evidence.

Since God is the author of the two books (Bible and nature), we must accord them the same level of integrity. They both are infallible. So, when we find apparent contradictions in the Bible, it could be because we do not understand the scripture or are misinterpreting it.

We need then to dig deeper for revelation and the right understanding of the issue. Similarly, when science and the Bible have apparent contradictions, it could be that our understanding may be limited, or we may be misinterpreting one or both books.

I recently read an interesting opinion on why science cannot answer "why" questions. Lars Kristan suggested that the word "why" per se is imprecise and can have two meanings a) is "why" asking about the cause (or reason), OR b) is it asking about the purpose (or intent)? Science can answer the first but not the second. It is not the place of science to search for meaning or intent where there is none. Doing so will cause science to slip into religion's territory, degenerate into an agent of conspiracy theories, and become pseudoscience. Caution: Let each stay in their lane. Let us leave the second "why" to priests and prophets. By the same token, pastors, please think twice before you demonize science, for it is the only method by which we can decipher God's book of nature. When scientific information (objective truth) appears to contradict the Bible (personal truth), don't panic and, out of ignorance, declare it as coming from the pit of hell, as some do.

What we don't understand now in science could be explained or discovered in the future. There is a theological perspective called the God of the gaps, which some use to explain or attempt to prove the existence of God.

Simply, this perspective posits that if scientists cannot explain it (i.e., there is a gap in their knowledge), then God takes over. But what happens when scientists find the missing information to fill in the gap later on? Does God then no longer exist? This perspective is obviously fallacious.

The stickiest point between science and faith is the origin of the universe. Christians of all ages have pondered or asked questions like Who created the universe? If God created the universe, who created Him? Why did God create the universe the way He did? Why are we here on Earth? It is okay for Christians to say God created the universe and quote from the Bible (Genesis 1:1). God has no beginning and no end; He has always existed. He created all things for a purpose. There is a purpose for everything under the heavens, states Scripture. So, for the people of faith, the question is not how but why are we here? What is our purpose as humans on Earth? Is there life after death?

If so, where do we go after death? Science cannot answer any of these questions. Science is interested in knowing, for example, what keeps the celestial bodies suspended in space. How did it all start? The current theory of how the universe came into being, the Big Bang theory, suggests that both time and space emerged together over 13.8 billion years ago, along with energy and matter, and with this came the laws of physics. But where did energy and matter come from? What happened immediately after the Big Bang? In fact, the universe, according to Hubble (as in the Hubble telescope), is not static but expanding. Is creation an ongoing process then, some ask? According to scientists, about half a million years after the Big Bang, the cosmos began to cool. The subatomic particles began to slow down and come together, a process that repeated itself over eons of time to form the structures of the universe. This is all about the heavens, the domain of astrophysicists, cosmologists, theoretical physicists, and similar professionals. The earth is part of the planetary system of the universe, located in the galaxy called the Milky Way.

In addition to telescopes, some of the latest breakthroughs in the science of the universe have come about because of the most complex machine ever built by humans the Large Hadron Collider (LHC) located in Switzerland (CERN). This immense structure, technically called a particle accelerator (accelerates a group of particles called hadrons), or casually as an atom smasher, allows scientists to accelerate particles in two beams in opposite directions close to the speed of light, through 17 miles of a tube, in ultra high vacuum, so they smash into each other and break into subatomic particles. Some of the critical questions scientists hope to use the LHC to answer include the origin of mass (why some particles are heavy while others have no mass at all) and what are dark matter and energy? In 2012, the existence of the Higgs Boson (nicknamed God particle) was confirmed by scientists working with the LHC, helping to address the question of mass. Advances in astrophysics and physics reveal new information to the effect that the universe has a beginning. This is music to the ears of Christians who pounced on it to say, we told you so, in the beginning, was God. The bigger message is that there is no need to bash science from the pulpit with ignorance and insults. Thank God for STEM! It is helping us better understand nature.

One of the subjects on which religion has been at loggerheads with science for the longest time is the origin of life. First, Christians believe that humans were created by God, while scientists believe that all life forms descended from a common ancestral cell (called the Last Universal Common Ancestor). Another way of saying this is that all life on Earth is related and connected. The Theory of Evolution, proposed by Charles Darwin in 1859, is the reigning scientific explanation for how the diversity of life emerged on Earth. Darwin summarized evolution as descent with modification, whereby species change over time and give rise to new ones that share a common ancestor by a mechanism called natural selection. Evolution's fundamental principles are variation, inheritance, selection, and time. Individuals of a species are not identical but vary in traits. This is the result of mutations or sexual reproduction. Because of limited natural resources, organisms with heritable traits that favor survival and reproduction tend to compete better and leave more offspring than their counterparts (survival of the fittest). The favored traits become adapted to the environment over time. Like him or not, Charles Darwin was one of the most brilliant minds that walked the earth. His work was simply genius, mainly because he did all that monumental work without benefiting from the modern science of genetics and molecular biology. However, both disciplines have since confirmed his findings.

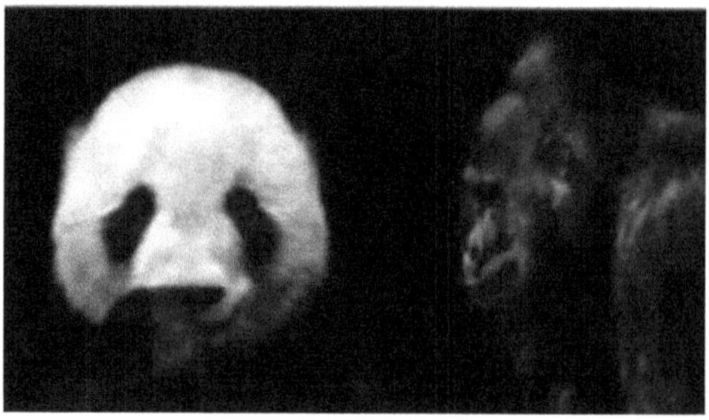

Evolution is currently held as both a fact and a theory. Natural selection can cause a species to change in small ways (e.g., color, height, size) over several generations. This is called microevolution and is incontrovertible even to most religious people who understand the subject, for it can occur literally right before our eyes. We routinely use the principles of evolution in agriculture, medicine, and biological research and applications. This is what plant geneticists and breeders (animal geneticists and breeders) like me do for a living (we are sometimes called applied evolutionists). The big difference between Darwin and breeders is that artificial selection, not natural selection, is the arbiter in modern breeding.We can create a new variety in about ten or so years versus Darwin's millions of years by using artificial selection. So, when you go to the grocery store for produce or meat, or go to the pet store to buy puppies, etc., thank God for Charles Darwin's pioneering work. The origin of life remains a theory that is actively being researched. This kind of evolution is called macroevolution (the evolution of one distinct type of organism into another), for which no empirical evidence exists (no one has seen it). Realizing that we cannot dismiss evolution, at least not in toto, some Christian scientists have proposed that maybe God created the universe by evolutionary principles or processes.

Scientific fact is no respecter of persons and will remain true whether you believe it. If you don't believe in the law of gravity, it does not negate its existence. But you are bound to be a believer in a hurry when you jump from a rooftop. As a Christian, I believe in the creation of the universe by God. I also believe in scientific facts about nature (i.e., creation). When the two systems of belief appear to contradict each other, I accept that there is so much in the Bible and science that we don't know. While nothing can make me abandon my faith, I will not disregard scientific truths. I believe that STEM is part of God's plan for His creation. We should believe in religious truth (personal) and scientific truth (universal). When I peer into a microscope (made possible by STEM) to view a biological specimen (of nature at work) on a slide and see the fantastic features before my eyes, I cannot but burst into the worship of how wonderful and great God is. My understanding and practice of science strengthen my faith in God which helps me better understand and appreciate the wonders of His work. As a scientist, I concur with the Biblical declaration that we are indeed fearfully and wonderfully made. Otherwise, how could humans accomplish such amazing feats in the world?

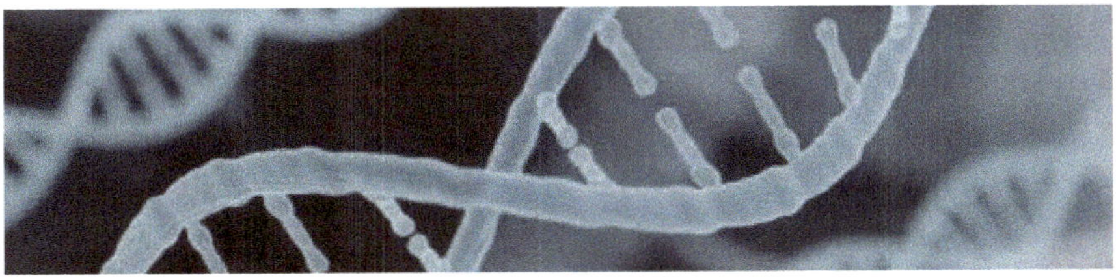

Some also suggest that DNA is sacred (connected to God and so deserving of veneration). The expression "we have God's DNA" is popular with some preachers and Christians who want to sound contemporary and in step with science. If not adequately explained, the problem is that it can cause more harm than good. Humans are both spiritual and physical beings. We engage God through our spirit, which we received at creation because God is spirit. We received His spirit when He created us in His image and likeness.

So, it is reasonable to say we have His "spiritual DNA." It establishes a relationship with the divine, God, something no other living being enjoys. This belongs to and returns to God when we die. This "DNA" is sacred. Our physical part (body) is governed by our physical DNA, which is unique for each individual. When we die, it returns to dust, it is not sacred. Comparative genetic analysis has shown that humans share about 96% of their DNA with the chimpanzee, 60% with the banana, and 60% with the chicken. God shares His "DNA"(spiritual) only with humans, not animals and plants. We are more than our physical DNA, though.

We were created as body (physical, material) and spirit and soul (immaterial). When Adam and Eve sinned, our spirit died. Now, through Christ, it can be renewed. It is agreed by scientists and others in society that there should be a limit to what we can do when it comes to research with humans. We should not manipulate the germline to create the so called "designer babies." Humans are a special creation. This is why many oppose the cloning of humans because human life has a connection with God, the sacred, through our spirit. We respect and value human life in a way that is different from how we treat other lives.

Misinformation and Disinformation

Another significant distinction between faith and science is that the faithful believe in miracles (events that natural or scientific laws cannot explain). Such events are attributed to divine origin.

Unfortunately, scientists cannot operate on this premise, whereby something can even arise from nothing. Scientists must start from something to create something new or better. Prayer is the practice in Christianity whereby we seek divine intervention or assistance in our affairs.One area worthy of discussion is healing. As a Christian, I believe God can heal me of any disease miraculously, but I also believe in medicine, the route by which He most often accomplishes healing these days. I could demonstrate this by doing something similar to what I did in Chapter Two, where I demonstrated unequivocally that STEM is part of God's plan for humanity. In fact, the behavior of some in the Christian community during the COVID-19 pandemic prompted me to launch this book without further delay. Many more lives could have been saved if we had listened to the advice of knowledgeable scientists instead of playing along with self serving politicians, aided by misguided religious folks and others to spread misinformation and disinformation.

I have seen several documentaries about attempts to provide scientific explanations for Biblical events, such as the ten miracles of the plagues visited upon the Egyptians for the Pharaoh's refusal to let God's people, the Israelites, go from bondage, as he commanded.

The conclusions are invariable that natural events were implicated in the events and were not unexpected. Miracles are events that defy human explanation. Something can arise out of nothing.

I personally believe that all things cannot be scientifically explained, no matter how hard we try. I am painfully aware that some unscrupulous religious leaders stage fake events to convince the audience that they possess miraculous powers. Some of their followers suffer needlessly or even die from false teaching.

We have a "specialization" (for lack of a better word) in Christianity called faith healing and practitioners who are called "faith healers." Some Christians feel that taking medicine is unchristian and displays a lack of faith in God. These same people will avail themselves of all that STEM has afforded us in society. So, what is so special about healing that we should suffer needlessly for lack of knowledge? Part of this behavior is due to wrong teaching by church leaders. God gave us science and medicine, and we must be thankful and use them when needed. I am not saying faith healing is a thing of the past. Absolutely not. But all modern global health challenges (e.g., polio, measles, and pandemics) were brought under control or eradicated through the intervention of science.

I ask you, how many people do you know in a day who receive miraculous healing or are raised from the dead? In medicine, a patient flatlines when the heart shows no electrical activity or appropriate signals on the monitor. The heart stops beating. This patient is near death or has recently deceased. Through STEM, we have devices (defibrillators) that can be used to electrically shock the heart back to normal. For all intents and purposes, dead or near-death patients can be brought back to life, an event that occurs fairly commonly and daily worldwide. Modern medicine is STEM based.

Let us thank God for science (STEM), His plan, and His gift to contemporary society. When we use medicine, we acknowledge God, who created intelligent humans and endowed them with the tremendous creative capacity to perform what we sometimes refer to as "medical miracles." This makes Him happy.

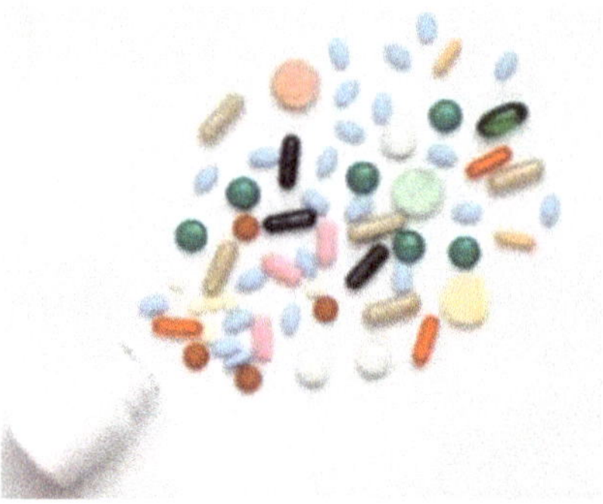

People can now survive diseases and injuries that heretofore spelled a death sentence for the patient. I am a survivor of prostate cancer. Thank God. I did not know I had cancer until at an annual checkup when my physician read my lab tests and noted that I had an elevated PSA level. A further detailed exam, followed by a biopsy of tissue samples from the prostate, confirmed the presence of the killer disease that takes the lives of numerous people annually. I am alive because of the prayers from around the world, which set the medical treatment in motion, involving sophisticated technology and medicines. But equipment or not, things could have been different if I had ignored the prompting of the Holy Spirit in me (what we sometimes acknowledge as "something told me") to schedule a checkup when I did. Life expectancy for many debilitating diseases has dramatically increased worldwide because of STEM. Let us thank God for his wonderful gift of doctors and STEM. Where doctors fail to affect healing, prayers can turn things around. Miracles do happen, still. There is a popular saying among Christians to the effect that "God heals, and the doctors take the credit." I don't think this is an accurate or fair statement. God routinely accomplishes His purposes through humans.

Another issue of concern in society is how scientists conduct their business. We often hear the accusation of "playing God" leveled against scientists when they pursue certain projects. From the theological perspective, God has reserved certain matters for solely divine determination, and such prerogatives must not be infringed upon or usurped by human beings.

The most common divine decision pertains to the end of human life. Since humans cannot create life, they have no authority to destroy it (human life is sacred - "sanctity of life" idea). Only God may end human life. Some ask if doctors and scientists err when they intervene to cure life-threatening diseases and prevent death through "medical miracles." Part of the basis for leveling the "play God" accusation is that God is omniscient and almighty. He knows the end from the beginning. Humans are very limited in this capacity to see into the future and hence must tread cautiously in their actions that impact life. But do scientists play God or imitate Him? If God is the creator of all things natural, and we are made in His image and likeness, is it any wonder that humans would be endowed with creativity and a flair for innovation? However, a note of caution to scientists is that just because it can be done does not mean we should do it. I mentioned earlier the universal opposition to manipulating the germline of humans. We serve society and should respect members concerns about our work.

Nature is God's classroom for scientists to learn the basic principles and concepts that govern how the universe works. After understanding the properties of matter and the natural laws in operation, scientists can manipulate nature to create new products or improve existing ones. Some of the critical tools in the biological toolkit of scientists originated from nature. They learn from nature and then refine those natural strategies into techniques or tools to make them more effective and efficient. One technique that is a significant part of biotechnology, and comes from studying nature, is genetic modification, whereby a foreign gene (DNA) is transferred from one organism to another. A widely occurring soil bacterium, Agrobacterium tumefaciens, causes crown gall disease (a tumor) in some plant species. It does so by an ingenious strategy whereby it inserts its genetic material (animal DNA) into the cell (plant DNA). The bacterium possesses a circular DNA (plasmid) that contains a tumor-inducing (called the Ti plasmid) segment, which is transferred to the host cell upon infection. In the host cell, it transfers a segment of the plasmid DNA called T-DNA (transfer DNA) into the chromosomal DNA of plant cells.

This process causes the hijacking of the cell division mechanism of the plant cell to produce tumor cells. Scientists studied this mechanism and adapted it for transferring alien or desired genes into plants. First, they locate the tumor causing part of the DNA and replace it with the desired DNA (gene of interest). The bacterium with the recombinant plasmid (with the alien DNA) is said to have been transformed. The plant of interest is deliberately infected with the transformed bacterium, allowing the bacterium to proceed with what it does naturally. Only this time, it has been tricked into being a courier (vector) for a good gene, not a tumor causing gene. The process is called Agrobacterium transformation and is a very popular tool in plant genetic engineering. Some in society cite religious concerns about biotechnology and how it infringes on God's prerogatives (playing God) and uses unnatural technologies for genetic manipulation. Well, as you see, scientists copied what nature does and made it better. If the Agrobacterium is a naturally occurring microorganism, and God created nature, He must have created such a survival mechanism for these microbes in natural settings. God, by extension, must approve of the mechanism that scientists in genetic engineering widely imitate. Scientists use this mechanism for good, not to cause diseases, but to prevent them and bring about cures.

Part of the issue with this technology for some believers is mixing human DNA (which they deem sacred; it is not) with a foreign one (bacterial). Another highly successful application of genetic engineering is the commercial production of insulin for use in humans. Previously, insulin was harvested from the pancreas of pigs and cows. The drug was very expensive to produce. These days, the gene for insulin is extracted from humans and inserted into the plasmid DNA of the common yeast, Saccharomyces cerevisiae (or E.coli). The transformed bacterium is grown in fermentation tanks for it to multiply. The material from the tank is harvested and processed to produce purified insulin. This process has received little public protestation, if at all, even though human DNA has been joined with bacterial DNA. Diabetes plagues about 6% of the world's population. In plant breeding, USDA records show that in 2019, about 85-95% of all corn varieties planted in the US were GM designed. Some were only for herbicide-tolerant varieties, while others were for multiple traits (stacked gene varieties), e.g.herbicide-tolerant and pest-resistant.

The percentage for soybean was even much higher, ranging between 91 - 99%, depending on the state. This means that GM is generally accepted in the US after years of intense public protestation. On the animal breeding front, a genetically engineered product has been applied to boost milk production in cows. Cows produce a growth hormone called bovine somatotropin (bST). Scientists discovered the gene for this protein and genetically engineered a version of it using the recombinant DNA technique called recombinant bovine growth hormone (rBGH). The commercially produced hormone is marketed as Prosilac™. This hormone boosts milk yield significantly. FDA approved its use in 1993. But consumer outcry has led to many dairy companies ditching this biotech product. Currently, less than 10% of companies use the drug in milk production.

Back to a sentiment in society that I broached earlier in this chapter. When scientists announce an outstanding achievement, especially when the method used is controversial, people question whether we are "playing God." This can be interpreted in one of two ways, positive or negative. When a scientist plays God, it could be construed to mean the achievement is so mind-boggling that it borders on the supernatural. Only a person with divine endowments could achieve such a feat. But when we recall that humans are made in the likeness of God and hence are endowed with creativity and ingenuity, then we can also say it is not surprising, for the apple does not fall far from the tree. Like father, like son or daughter. As noted earlier, we often use "the miracles of modern medicine" to acknowledge outstanding feats achieved by doctors and scientists. Doctors and other healthcare practitioners, for all practical purposes, can raise people from the dead (or near-death) by applying medical technology.

However, more often than not, the expression is interpreted to mean scientists have gone too far. Who do they think they are? God? God is all powerful and all knowing. He knows the end from the beginning. Scientists do not know beyond today with certainty. We don't know if our inventions, like a new crop variety, a so called GMO, will exhibit adverse health effects in the future.

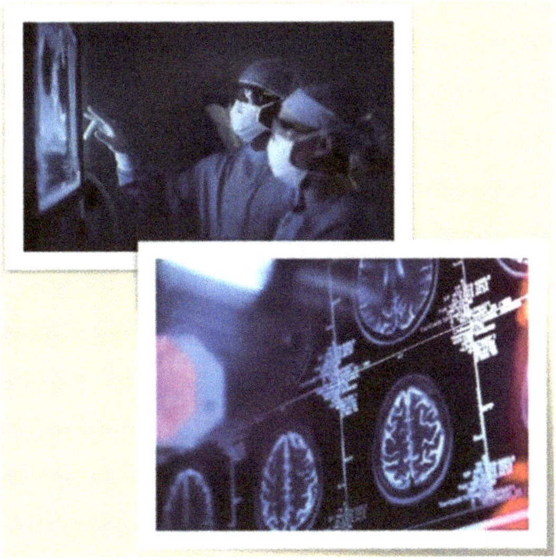

Research shows that religious people generally are slow to adopt technological innovations. I know from experience that modern musical instruments were unwelcomed in the church as instruments fit for use in worship. The organ was the only standard church instrument for centuries. Drums, guitars, synthesizers, etc., which were associated with the rock and roll genre of music, were mainly frowned upon. Fast forward to today, and you'll find some of the church music and instruments that can rival the best rock concerts. The reason that is commonly invoked to account for this trend is that contemporary music draws the youth to church.

In my opinion, there is no reason for science and religion to be controversial. The methods of discovering truth in these domains are diametrically opposed to each other. As I once stated in this book, we go up to heaven by faith, but we go up to space by science.

In 2020, the world watched as Elon Musk's Space X vehicle launched into space with two astronauts on board and successfully docked with the International Space Station, which is suspended some 250 miles above the earth. This spectacular achievement was the result of humans applying STEM. The launch was only sanctioned when all systems were a go, as it is said in the space industry. There were many test flights before the actual flight on that day. When a weakness in the craft was discovered, the problem was fixed by the appropriate engineers and technicians. Even on the launch day, the flight was postponed when conditions were not conducive at the eleventh hour. NASA did not brush it off and call priests to the chapel to pray for the problems to disappear. Science cannot operate by faith, only by sight. If things don't add up, we cannot pray it away. This is not to say that some of the scientists and technicians who are Christians did not pray to God for wisdom to address problems. Christians believe that God can take it further beyond human capacity.

I personally believe that once we understand how faith and science work and accept that each has its own set of rules under which they operate for the good of society, both systems of truth can peacefully coexist in society.

As a Christian who is an educator and a scientist, I believe the only way to teach science in schools is as science and according to its language and culture. Any other method will be disingenuous to students. The battle in public schools about evolution is unnecessary and emotionally charged. Until Intelligent Design is submitted through the traditional method of science process and found worthy, it does not belong in science textbooks or school science curricula. Accepting it as science when it has not passed the standard test is nothing short of intellectual fraud. It will accomplish nothing beyond deceiving students and satisfying the egos of religious adults and political allies. As an educator and Christian, I will teach creation at Sunday School, but never in the biology classroom, and at the end of the day, I thank God with a grateful heart for both. After all, science is about discovering and understanding nature, which God created.

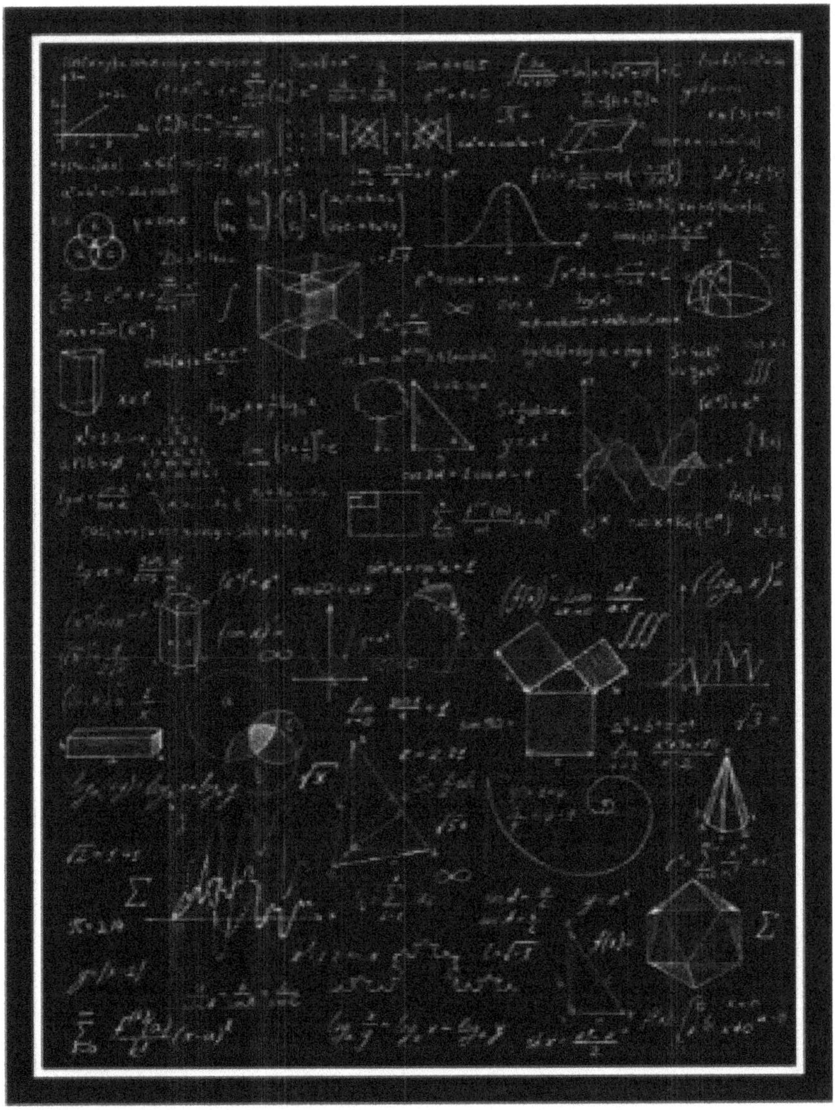

CHAPTER 10

HOW POLITICS WORKS

POLITICS

Politics impacts every aspect of our lives. It affects how science (and STEM in general) is supported, conducted, and deployed in society. Politicians, generally, don't have the best image in society. Well, they have themselves to blame, for actions speak louder than words. We often hear or use the phrase "playing politics" that connotes acting for political gains rather than on principle. But the fact is that people or society make politicians, for we vote them into office and nurture their behavior for as long as it serves our purpose. In this chapter, I discuss the world of politics and politicians enough for us to manage our expectations of the latter group.

According to the dictionary, politics refers to the activities associated with the governance of a specified jurisdiction of society (e.g., state, country). The governance follows a structured system for making official and binding decisions on the citizenry or the governed. The system of government distributes power among different parts and levels of the jurisdiction. The decisions made by the governing authority include how to use, produce, and allocate resources and the well being of society (e.g., security, health, economics, law and order, justice, equality, etc.). The discussion in this chapter is only introductory to the subject of politics. It is designed to provide a brief overview that would allow its impact on STEM in society to be effectively analyzed.

Various political or governing operate around the globe; the most common that we hear or read about include democracy, republic, dictatorship, monarchy, theocracy, totalitarian, authoritarian, communism, socialism, federation, and anarchy. These are not tightly packaged or exclusive systems of governance. A nation may officially combine two or more types at one time or be forced by circumstances (e.g., civic unrest, war, coup d'etat) into an unplanned system of rule. For example, the US combines democracy and republic and embraces socialist concepts when appropriate and advantageous. Further, a leader may start with an established governance style or philosophy, like democracy, and veer off track to embrace communism and dictatorship, wholly or partially, when convenient.

A democracy is generally described as a government of the people, by the people, and for the people. This simply means that anyone in the citizenry may run for public office. In a republic, people are elected into the government, but it is headed by one person who the people also elect. A theocracy is a form of government where the dominant religious institution holds power over the ruler. The nation is governed according to religious doctrine.

You may have heard about the "Sharia laws" in some Islamic nations. Ancient Israel, God's chosen people, was for a while a theocracy, where God was their ruler. Then they lived under judges and prophets who communicated God's will via the command, "Thus says the Lord." Modern Israel operates a parliamentary democracy with a prime minister and a multiparty system. A king or queen governs a monarchy; the position is hereditary and was vested with absolute power in the olden days. Some modern monarchies operate within democracy (UK), with an elected prime minister, thereby reducing the monarchy to a ceremonial role. I will proceed with simple dictionary-based definitions of some of these political systems.

A dictatorship arises when an individual gains power by force (e.g., coup d'etat) and imposes policies on the people. To gain the international community's approval, some dictators try to act in the interest of the people (the so called benevolent dictators) and less out of self-interest. An apt description of a totalitarian government is a dictatorship by a party instead of an individual. The country is ruled by one party or group of people mainly for their self interests. No country sets out to be ruled by anarchy. Unfortunately, civil unrest and competing ideologies can plunge a country into a state where the elected government is paralyzed and nonfunctioning. Communism, on the other hand, is a revolutionary socialist movement that sets out to create a society in which class and private ownership are abolished or suppressed. The community owns the property, and the government controls all activity. A federation is a collection of small states that govern their internal affairs but submit to a central government.

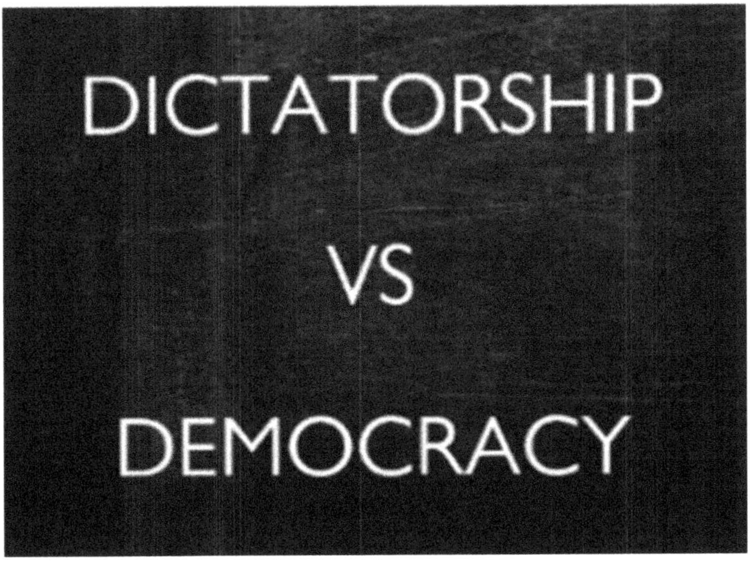

There are pros and cons to these systems of government. Eventually, it boils down to the leader in charge and provisions available to ensure that it works as it is supposed to. Democracy perhaps offers the most freedom for the citizenry to participate in the governance and affairs of the state or nation.

People are free to express their opinions, protest the mal-administration of the government in power, and remove leaders from office for incompetence or violations of laws. No one is above the law. But as I indicate later in this chapter, human beings will always find a way to make a good thing bad. Some democratic leaders are self absorbed and lead not for the benefit of the people but instead for themselves.

I will focus on democracy and republic, systems of government that I am most familiar with, and reference the US when appropriate. As necessary, examples will mainly be drawn from the last decade (the Obama, Trump, and early part of the Biden Administrations) to make it meaningful. As a federal republic, the US system invests power in three coequal branches of government the Executive (President), Congress, and the Judiciary. The current constituent fifty states govern their own internal affairs under an elected governor. Congress consists of two chambers at the federal level, the Senate and the House of Representatives. The Democratic Party (DNC) and the Republican Party (RNC, popularly, GOP) are two major political parties. These parties have their audiences or "bases" to whom they cater or pander with their message. Traditionally, the DNC is known for representing the interests of working families and fighting for justice and equal opportunities for all Americans. The RNC focuses on individual liberty, free trade, and free markets. It was founded to stop the expansion of slavery. The RNC's platform is based on the ideology of conservatism, while the DNC's platform is based on liberalism. The RNC advocates for a strong defense, small federal government, and minimal regulations on business and promoting of personal wealth.

The DNC advocates for a centralized government solution to problems, such as education, health care, social security, and welfare.

These are representative goals of these two parties. Having lived in the US for nearly 40 years at the time of launching this book, I know that the positions of these parties shift from time to time. I was surprised to read that the RNC was founded to stop the expansion of slavery, while the DNC opposed this position at one time. However, these days, it appears the DNC is more active in addressing the needs of those who were disadvantaged by slavery and other social injustices. In contrast, the RNC sometimes appears even to oppose such efforts. Political parties "encroach" on each other's turfs for advantage (i.e., one party may adopt another's traditional position when advantageous). I once saw a cartoon that described a typical politician as one who stands for what others fall for.

I cannot agree more with this observation. There is no doubt that politics is "a game" requiring experience and tact to succeed. During presidential elections, candidates start by floating around possible slogans to see what will resonate with the public, especially their base. Every politician or political party claims to speak for "the American People." In my opinion, "America" may be politically defined as the number of people needed to win an election. Calculations are made to determine which states or social constituencies favor what the party or candidate stands for. Once they have ascertained a path to victory, all other Americans are secondary and don't matter, maybe until after the elections. The political base sometimes appears to be generationally supported. We sometimes hear the slogan. "Once a Republican, always a Republican."

STEM is changing how politics is conducted in this modern era. Polling is aided by our current information/ technological era, which provides easy access to information. Electronic voting machines facilitate the voting process, making it easier to tally results while minimizing human errors and intentional biases. All the polling and statistical analyses done by political teams are thanks to STEM, especially social and political sciences. The modern invention "Social Media" makes it easier to engage the citizenry for political purposes. There is such a thing as "opposition research" in politics, which is another name for finding out "dirt" on a political opponent (information that is damaging to an opponent's campaign).

Truth be told, we all want what is good for America, provided America means all Americans. The goals of the two parties, when combined, will make America a much better place for all. The parties must learn how to give and take (aka compromise). There have to be bipartisan approaches to finding solutions to problems in the nation. Of course, either party wants to be in the driver's seat to have the opportunity to push through specific ideological positions or initiatives. The GOP often labels some democratic leaders or candidates as socialists. But some of the most enduring privileges in America, such as Social Security, Medicare, and Medicaid, are essentially massive "socialist" programs enjoyed by all Americans. On balance, there are also massive and transformative projects undertaken by GOP leaders in the past, such as the federal highway system and NASA.

Some quality collaboration and compromise between the two political parties were common some decades ago. Unfortunately, the GOP and DNC relationship has soured significantly in the last decade, especially in the previous three presidential cycles, the Obama, Trump, and early Biden eras.

The election of the first Black President was hailed as a momentous occasion that was potentially transformative and expected to make racism in America a thing of the past. How naive we were! The baseless and vicious political tactic, the infamous "Birtherism attack," used by Mr. Donald Trump to undermine Mr. Barack Obama, failed to deny the latter's election to the presidency. But before he was even sworn in, the GOP, led by Mitch McConnell, declared emphatically that their number one goal as a party was to ensure that President Obama was a one time president. This was the first salvo in the political warfare between the two parties in the first season. Needless to say that bipartisanship went down hill, leading to Congress receiving some of the lowest approval ratings in history. President Obama, an astute politician in his own right, survived his first term and won a second one. Experts rank him among the most successful of modern day presidents.

The surprising ascension to the presidency by Mr. Trump in 2016 ushered in a new kind of politics in America. First, it appeared he was bent on pursuing some "unfinished business" and did not waste time attempting to undo everything President Obama accomplished (Obamacare, DACA, etc.).

But whereas that might be considered "politics as usual," it was his brand of divisive and "galvanize the base" politics that were problematic in his tenure in office.

President Trump was a master at playing base politics, like him or not. He famously declared that he could stand in the middle of Fifth Avenue in New York, shoot a person, and still not lose a vote. We should have believed him, for it played out over his four years in office. He did not hesitate to use every tactic possible, including stoking fear and provoking racial tensions, his trademark lying, and being quick to declare anything he did not like as fake or a hoax. I believe lying or stretching the truth a little is not unique to President Trump; he just made it a political art form or language. As is often said, his tenure in the office exposed the true America. He gave voice to people to vote or act out how they felt in their hearts but had been previously afraid to reveal in public. He was not afraid to exploit racism to his advantage and spoke in support of white supremacy.

Surprisingly, or maybe not, he successfully kept the White Evangelical vote in his corner, thereby effectively silencing the group that once claimed to be champions of morality in America. I agree it is not fair to say Trump silenced them, for they chose to keep silent when the voice of reason, truth, and compassion needed to be heard.

The other here to fore uncharted waters in modern US politics is foreign interference in our elections. The Mueller Report showed that Russia, especially, interfered in the 2016 presidential election in favor of Mr. Trump and perhaps was set to continue on this path in the future. The media is another critical asset in winning the base wars in modern politics. For example, the GOP has the stead fast support of FOX News, while the DNC can rely on MSNBC for political support. There are also the radio and TV stations and the various political talk shows that support either party. The GOP appears to have an advantage as most major secular and religious radio and TV stations favor them. The print media also have their loyalties; The Wall Street Journal is considered favorable to conservatives, while The New York Times favors the liberals. The global phenomenon called social media has emerged as a force to reckon with in politics. President Trump effectively utilized Twitter, and especially Facebook, to promote his brand of politics disinformation, misinformation, fear, and lies. These global communication platforms were fertile ground for foreign actors to subvert the US elections. On the religious front, people of faith tend to see their candidate for president as "anointed by God" and are willing to forgive their short comings by saying things like "we are not electing a saint; we are all sinners. "But their opponent is not accorded such privileges.

Some other political tactics in US politics unfairly tip or influence an election in favor of one party or influence decision making for political gains.

The first is gerrymandering, manipulating an electoral constituency's boundaries to favor and entrench one party or class. A ruling party in the state government can redraw the electoral map, a task usually conducted with partisan fervor. The result can be so outrageous that some political districts are set in stone for years in favor of one party, thus ensuring that the seats cannot be lost for a long time (called safe seats). Representatives from safe seats can afford to be vocal or refuse to toe the party line. The other political ploy widely accepted in society is lobbying, an action meant to influence a politician or public official on an issue. Some have turned this in toan art form or even a profession (lobbyists). It can include a planned and civil audience with selected influential politicians on behalf of a cause but can sometimes assume the form of vociferous activism via protests and demonstrations.

That said, some political tactics are deplorable and unacceptable in a democracy. Voter suppression, a deliberate attempt to discourage legitimate participation in the electoral process, is conducted in various forms in the US. According to experts, they are mainly conducted to suppress the level of participation by racial minorities, predominantly African, American, Hispanic, and Asian voters, whose involvement in the democratic process surged from 2004 to 2008. The ruling party at the state level can deliberately implement voter ID laws that discourage voting by these minority populations, who are known to support the DNC in large numbers. Other tactics include reducing early voting opportunities, eliminating polling stations, and purging voter rolls. In the 2020 elections, the systematic disenfranchisement tactics were unbecoming of a nation that portrays itself as a global beacon of democracy.

There are some nations in which refusing to vote when one is qualified to do so is a violation of the law. Following the loss in 2020, the GOP went on a nationwide crusade, introducing hundreds of bills in GOP governed states to disenfranchise minority voters.

Another deplorable tactic is delegitimizing mail in voting. Even though there was little, if any evidence, that such a method of voting is fraudulent, the Trump administration railed against it following the loss in 2020.

Other related political ploys used in American politics are "pivoting" during a political debate, "spinning the facts," "creating false equivalence," and "silence." Politicians evade questions by reframing them in ways that allow them to focus on their well rehearsed "political talking points." Following a political debate, politicians and their surrogates make the rounds of the news media circuit to spin the outcomes in their favor, often visiting their favorite TV and radio talk shows. They use the opportunity sometimes to attempt to "snatch victory from the laws of defeat" and correct misstatements made by their candidates. The false equivalence tactics are used to avoid admitting guilt or addressing a problem head on. A contemporary example is the 2020 global chant "Black Lives Matter (BLM)." This slogan emerged from the need to address police brutality in America, whereby African-Americans are systematically abused by the police, sometimes resulting in the death of victims. The message is clear and very simple. The US Constitution declares that all men are created equal. If that is the case, the BLM movement insists that all men should be treated equally in all aspects of society.

Unfortunately, social science research shows that African-Americans are anything but equally treated in the justice system on a systematic basis. If all lives are equal and matter, the BLM movement demands that Blacks be treated as if they are part of the "all." To avoid tackling the issue head-on, some say, well, the police are also sometimes victims at the hands of criminals. Therefore, Blue Lives Matter should be touted as well. But this is a sad cop-out. Losing a law enforcement officer in the line of duty is deplorable and must be prevented. But the truth is that Blacks don't systematically seek out the police to attack them. Instead, the police are charged with systematically profiling and confronting Blacks (on appearance, mere suspicion of criminal activity, minor or no infraction of the law, stop and frisk laws, etc.). These events often result in senseless deaths of victims. The cry of the Black Lives Matter movement is for such injustice to stop. Citing the killing of police to excuse inaction regarding the plight of minorities in the justice system is pathetic and tantamount to dereliction of duty by elected officials, who must know better to care for ALL citizens without bias.

When in office, it is expected that the president and his party will do some political favors for those who put them in power. Some appointments are purely political and done at the discretion or whim of the president, while others need congressional approval. Even in the latter case, appointees are confirmed by votes that, more often than not, are cast purely along political party lines. The GOP will select persons from their conservative base, while Democrats choose persons with liberal views. Appointees are expected to toe political lines. As long as appointees are reasonably qualified, they can be accepted by the public. Unfortunately, there are times when square pegs are forced into round holes. Also, there are times when appointees go overboard in their loyalty to the president, as was the case in the appointment of Attorney General William Barr, whom some accused of acting like he was the personal lawyer for President Trump instead of the people's lawyer. The Trump Administration was also characterized by frequent resignations and terminations, some of which appeared retaliatory for speaking up against the president.

Politicians in Congress have the authority to summon federal political appointees to appear before the many committees they run in their roles as policymakers. They also have the authority even to investigate the president, should there be evidence of misconduct. Called an impeachment, it is a process by which a legislative body brings charges against a government official. It is equivalent to bringing an indictment against someone in a criminal proceeding. However, the process cannot force an official out of office. Congress successfully impeached Attorney General Eric Holder (Obama Administration) and President Donald Trump.

Another powerful provision under the democratic rule is the freedom of lawful assembly and to protest. More often than not, such rights are exercised without a problem.

Sometimes, people organize counter-protests to confront opponents, a recipe for disaster. In some cases where the public feels a court decision is unfair, peaceful protests sometimes degenerate into riots when unruly individuals take advantage of the situation to pursue destructive paths (looting burning stores, etc.). Occasionally, unfortunately, law enforcement, in its desire to quell a riot, inflames it by excessive force, resulting in death. Peaceful marches in America have been known to yield great dividends when they force the hand of government to enact fair laws or reverse previous unjust ones. Peaceful protests characterized the Civil Rights Movement that resulted in enacting laws for a more equitable society in America. Despite the shortcomings of the US democratic process, at least it can be said that, when the dust settles, change in leadership would have been safely conducted at the ballot box, usually without much drama.

This appears to have been the norm until President Trump lost reelection in 2020 and decided not to go quietly.

After losing over 60 court cases seeking to reverse the outcome of the elections, he urged his followers to believe in the lie that the election was stolen, the so-called Big Lie. This cumulated in one of the indelible stains on democracy in America when his followers stormed the US Capitol Hill on January 6th, 2021, while Congress was in session to certify the voting results. This was the lowest point in American democracy.

A violent takeover of the government, as occurs in some developing countries, is not part of the American political experience. But this does not excuse the need for much improvement, something that can happen if only the voters who elect the lawmakers will demand such changes. President Obama aptly put it, "Don't boo, vote." Politicians know that the bark of the electorate is louder than their bite. Many vocal citizens about social issues do not show up on Election Day to cast their votes.

Consequently, politicians are left with mobilizing their base and counting on their party's die-hards to show up. They use all forms of tactics, clean and dirty, "to energize the base," as they say, and get them to the polling station. It goes without saying that we get the leadership we deserve as a nation, for we are the ones who elect our officials.

Another characteristic of politics is that every administration (president) wants to leave a legacy or signature achievement. Some are fortunate to have their names attached to their legacy (e.g., Obamacare for President Obama).

Others, despite positive achievements, are forever associated with an unfortunate misstep that had disastrous consequences (e.g., President George Bush's Iraq War and President Jimmy Carter's Iran Contra War). William Shakespeare said it best when he said, "The evil that men do lives after them; the good is oft interred with their bones. "For political advantage, politicians will resurrect past mistakes that are of political advantage to smear their opponents. Some even will go further to dismantle the achievements of their predecessors deliberately. How low can some politicians go? It may not be low enough.

I believe most politicians are good people who happen to be playing the "game" of politics, in which players bent on winning can sometimes deny the truth. Yes, it is right, but if I admit it, my opponent will have a political advantage over me. And so, I deny or refrain from supporting a good proposal. Understanding this dynamic in the political arena will allow one to manage their expectations of politicians.

CHAPTER 11

SCIENCE VS. POLITICS

Science as a human endeavor can be expensive to pursue at a high level of excellence. Mendel made his seminal discoveries in heredity in a pea garden in the backyard of a monastery. But there is little of great consequence these days that can be discovered without a significant capital investment. To finally discover the elusive "God particle" (Higgs Boson) took the construction of a nine billion dollar LHC (Large Hadron Collider).

The federal government is the leader in funding research in the US, which means that lawmakers and politicians must approve of such spending. So, how does a domain devoted to objective truth interact with one that sometimes operates by convenient truth?

I indicated previously that STEM is the backbone of the economy of modern nations, and more so in this highly technological and information based society. The nations that invest in developing these disciplines will have a competitive edge over others in this race to control the global economy. America is known and respected globally for its outstanding leadership in science and technology education, research, innovation, and application. Students from all over the world flock to our higher education institutions to train in STEM. Our discoveries and technologies are coveted and even stolen, sometimes through well orchestrated sponsorship by other governments.

It is safe to say science and politics are practically joined at the hip in our modern society. Science is the pursuit of knowledge, knowledge is power, and power is politics (Elizabeth Lopatto).

Money is a significant factor in this relationship, but religion or faith is a strong influence as well, as will be discussed in a future chapter. Because the enterprise of science in the US is largely federally-funded, politicians sometimes play politics with it, to the detriment of its growth and advancement. As previously discussed in another chapter, there are two science categories basic and applied. The fundamental discoveries of nature that fuel technological innovations are in the domain of basic science. Because basic research is done without a specific application, its immediate value is not readily apparent to some decision-makers.

Consequently, it is an easy target for politicians for funding cuts. In fact, it would have been the US basking in the glory of the discovery of the 21st century, the Higgs Boson, which was eventually discovered at Switzerland's Large Hadron Collider (LHC) in 2012. But alas, politics happened, and the construction of the Superconducting Super Collider (SSC) in Texas, many times more powerful than the LHC, was nixed by the US Congress. Just like plant breeding depends on variation (as building blocks) for success, applied science and technological innovation cannot be advanced without basic research.

Scientists may decide on the research areas or topics to focus on. This is the prerogative of faculty members in academia. Such direction is influenced by factors including creativity, inquisitiveness, and genuine need in society. At least in the public domain, what gets done is mainly determined by politics, especially if it is capital intensive. In the US, federal funding (public funds) for basic research is administered through selected agencies, the major ones being the National Institutes of Health (NIH), Department of Defense (DoD), and the National Science Foundation (NSF). The funding allocation to federal agencies is decided by politicians, who may neglect to consult with scientists, and instead respond to the larger political community (voters, lobbyists).

As previously indicated, every political administration wants to leave a legacy or have a signature program. One leader may choose to emphasize STEM, whereas another may opt to be known for a strong defense. Funding cuts will target agencies of less interest or advantage to the administration's agenda, while funds are diverted to favored agencies. And even within the agencies, politicians may seek to champion specific or pet research areas (e.g., diabetes, cancer, cybersecurity, etc.). It is not uncommon to find that a politician who a disease has adversely impacted will champion the funding for research into it.

Political decisions not only impact what research is done but also what research will not be supported or banned. Some of the darkest moments in American history have been associated with the maltreatment of its socially disadvantaged populations.

Slavery takes the cake and is in a class by itself, but using the vulnerable in our population as lab rats is equally despicable.

In research, scientists developed the polio vaccine in part due to unethically acquiring and using HeLa (Henrietta Lacks) cells. Ms. Lacks was an African American woman.

Similarly, the Tuskegee Syphilis Study was denounced for denying treatment for the disease to afflicted African American males so that researchers could study the long term effects of the disease. Using people as test subjects without their consent is unethical. These days, the common practice is to solicit volunteers for human studies and receive proper approval before conducting studies that involve humans as test subjects. I discussed IRB (Institutional Review Board) in a previous chapter.

When political winds change, federal policies and research direction can also change. As previously discussed, one of the contentious research areas of recent times is stem cell research. The George W. Bush administration restricted stem research, a faith-based decision, through the President's Council on Bioethics that he created. The restrictions were relaxed under the Barack Obama administration, which eliminated the Council altogether.

The Environmental Protection Agency (EPA), created by the Nixon Administration, has especially been in the crosshairs of GOP administrations. The Donald Trump administration did not disappoint, as he wasted no time railing against climate science and basic research, slashing funding for some federal research agencies. EPA has been a thorn in the flesh of industries that generate pollutants that are a health hazard to humans and life in general. Toxic site cleanups are expensive, and so are emission controls, which increase production costs. Not surprisingly, leaders of profit driven businesses lobby politicians to loosen or eliminate regulations that would compel them to invest in pollution control practices. At least in the past, industries dumped pollutants on land and in water bodies with impunity. Often, the victims are the poor and socially disadvantaged.

It is also not hard to see why policymakers and scientists should work closely together. Scientists need politicians to support the discovery of knowledge (information), which they then use to make informed decisions for the benefit of the entire electorate. Social science research is indispensable in the making and implementing of sound social policies. Whereas some of our leaders have vigorously promoted and advanced the development of science, others have chosen to feign ignorance of its role as the foundation of our modern economy.

It is disheartening to see leaders and their political allies stand in willful denial of science, denouncing outstanding findings as hoaxes or fake, ignorant of how scientists arrive at objective truth. Sometimes, it appears truth does not matter in politics for as long as an action is politically advantageous. But before I create the impression that politicians are to blame all the time, I must hasten to add that scientists do get it wrong on occasion. Sometimes, the mis- or disinformation comes from corporations, as was the case in the tobacco wars. Despite the mounting scientific evidence showing that smoking tobacco causes cancer and other illnesses, the tobacco industry mounted a vigorous campaign to keep its business intact. Further, as humans, we view the world through a colored lens, as previously noted. Our biases do not allow us to accurately receive and use facts or information. The natural propensity is to dismiss unfavorable facts and embrace advantageous ones. Further, our value system, rooted in our faith, drives our decision-making and can make us choose our beliefs over facts.

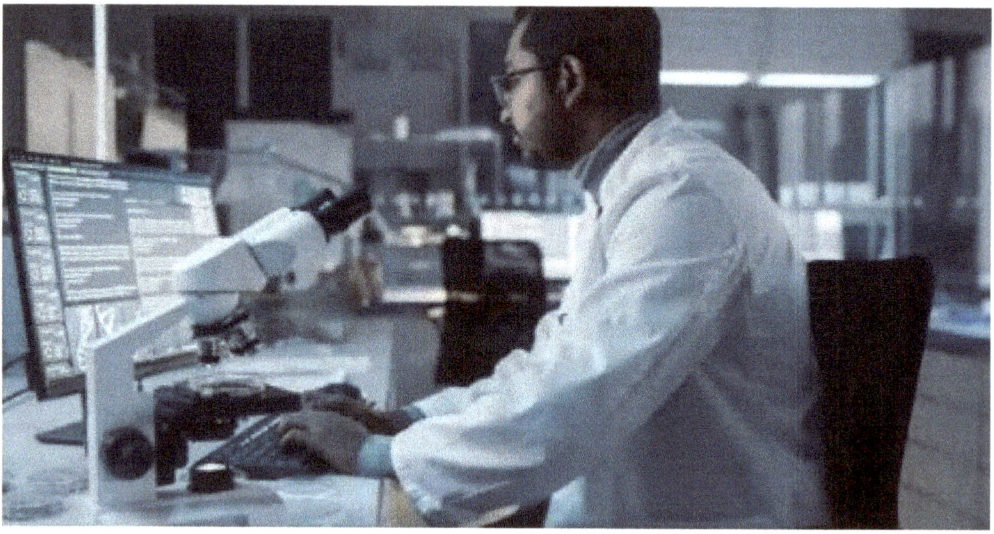

Because science thrives on objective truth, there is no opting out of whether or not to believe its findings. We do so at our peril. As the famed astrophysicist Neil deGrasse Tyson so eloquently surmised on one of his Star Talk programs, we must first accept the scientific facts to be able to make informed political decisions. Put another way; we must not put the horse before the cart, flippantly dismissing science. In a democracy, our politics can and often get in the way of proper and effective solutions because politicians are always looking for political advantage in their actions. It goes without saying that science and politics should be cozy bed fellows for our national security, economy, health, and the general well being of the citizenry.

What does it take to prepare for a career in politics? The US Congress is populated with persons from all backgrounds and levels of education law enforcement, medicine, law, military, science, education, political science, humanities, graduate degrees, etc.

The academic discipline of political science is devoted to studying both the theoretical and practical aspects of governments and is considered a social science. Some people enter the political arena on the wings of activism. But what about science? America needs to find and form STEM talent and expand minority participation in these disciplines to maintain our global leadership and competitiveness in STEM. Research shows that women and minorities are significantly underrepresented in STEM education and professions. For minorities like African Americans, the systemic racism in society deprived them of education in general, leaving them now to play catch up. Women are underrepresented in STEM largely because White males have dominated the fields. In short, for years, society has denied permission for Blacks, especially, and also women, to break into the White male-dominated fields. There is a need for this disparity to be corrected to diversify the STEM workforce.

A Pew Research Center study in 2014 sampled public opinion on 22 science related issues of concern to society. These were grouped into several categories as follows: climate change and energy, government funding, evolution, biomedical research, food safety, animal research, and space. The results showed that the political persuasion of Americans influenced their position on issues such as climate change and energy policy, and less so on food safety, biomedical research, and space travel. Race and generational differences were factors in some categories.

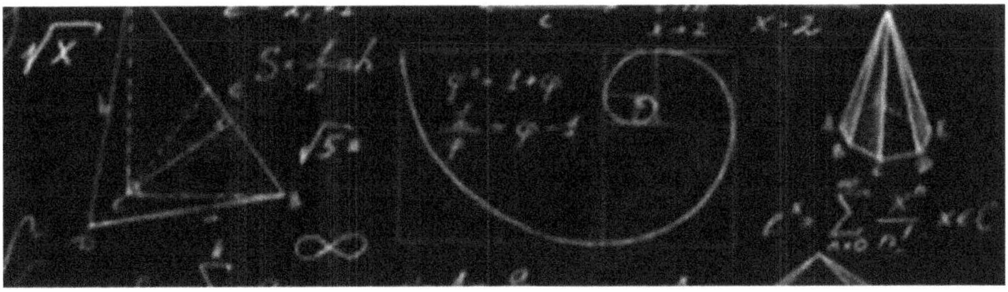

Let me take some time to provide a brief overview of how America handles STEM education and research. The US operates a very elaborate science education system and research enterprise. Youth training in science starts in kindergarten and goes through high school (K-12). As graduation requirements, states mandate a certain number of credit hours in mathematics and science at the pre-college levels.

In addition to the basic science education provided through science curricula, the Department of Education supports numerous programs promoting STEM. It is significant to note on its website that the agency is committed to ensuring that "A child's zip code should not determine their STEM fluency." There is no denying that research shows a disparity in the quality of public education in America.

These research institutes and agencies generate a massive amount of proprietary information (Intellectual Property) that must be protected from unlawful use. For this purpose, the US Patent and Trademark Office (USPTO) was established to serve the needs of inventors. Also, the Department of Homeland Security (DHS) ensures that students and scientists who come to the US to study or conduct research do so lawfully. We hear of graduate research students and even professors who attempt to steal and sell intellectual property to foreign governments every now and then. It is clear that efforts are being made to diversify the STEM workforce. Other areas of emphasis these days include computer science. It is clear that politics highly influence scientific research because it is funding intensive. Politicians must appreciate and support science that is not only applied but also theoretical. It is also critical that politicians listen to the recommendations of scientific experts rather than activists and lobbyists.

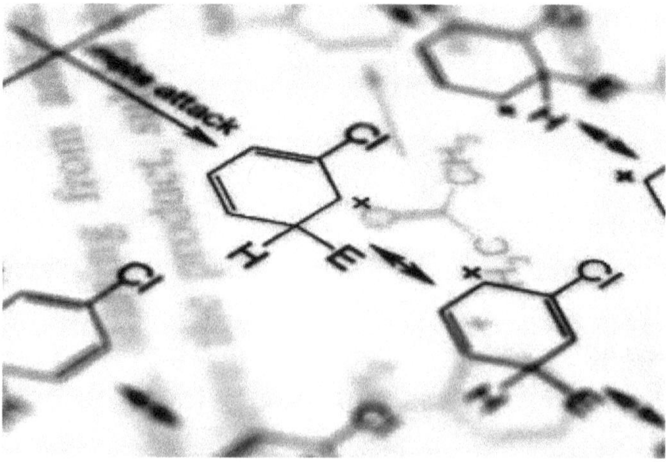

CHAPTER 12

CHOICE, ETHICS, MORALITY, AND STEM

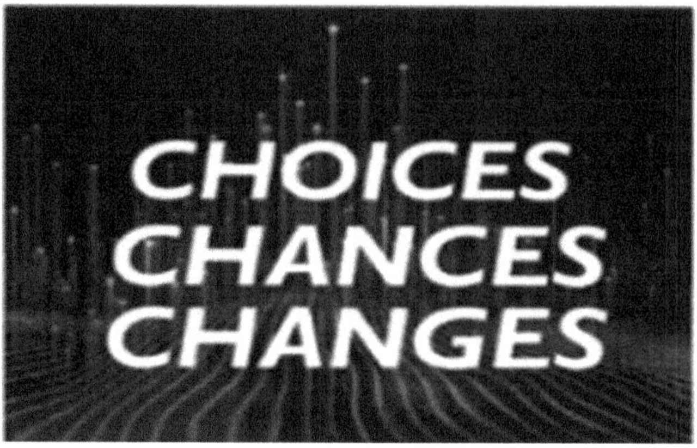

In previous chapters, I discussed the pervasiveness of STEM in society. Biologically, humans are unique individuals; no two are genetically a like (pick up mutations in the womb). One other unique aspect of humans is that we are divinely endowed with free will (choice). Through the practice of our faith and politics, our choices impact the development and deployment of STEM in society. Similarly, morality and ethics in society influence people's actions, including how science is embraced and advanced in society.

The three C's of life are choices, chances, and changes (Zig Ziglar). Life is one big series of choice after choice and decision after decision. Opportunity will knock, and you will have the choice to decide whether to let it pass or pursue it. Life is not static but constantly changing,day by day, both within us and around us. A famous adage, "Carpediem,"translates to "seize the day." Time and tide wait for no man (Geoffrey Chaucer). Opportunities are not available all the time, so make hay while the sun shines, yet another adage. We go through phases in life, each with its opportunities and challenges. Youth has its opportunities and challenges, and so does old age. Some things are best done and enjoyed while you are young. When confronted with choices or chances in life, you can "give up, give in, or give it your all" (Charleston Parker). This is another way of saying that you make life what it is or get out of life what you put in.

We cannot overemphasize the importance of free will or choice in life. From the sun's rising to the going down of the same, we make one decision or choice after another. We make small and big decisions and choose between options.

Our decisions or choices have consequences, good or bad. Even just turning left while walking or driving down the street instead of right can have life changing outcomes for better or for worse. The consequences of our choices may impact us exclusively and for a moment or a lifetime. However, they can also affect our immediate family or relations for a moment or even extend to generations. This means we must be careful how we choose between options. Christians believe God knows the end from the beginning, so it is wise to include Him in our decision making process and not ask Him to bless the plans that we have made on our own. Better still, He has the master plan for our lives, so we must work with Him.

I indicated in Chapter Three that God is the originator of choice. In His creation plan, He created humans to have free will to make their own choices in life and not operate like robots that an owner manipulates. If God ordained free will or choice, it must be good, for when he finished His creation, He declared it good and then rested. But if presented with options to choose from, the implication is that one could make a good choice when a better one is available or make a bad choice outright. God made it abundantly clear from the start that options have consequences, as seen in His warning to Adam and Eve that eating from the Tree of Life would result in death. You would think that such a stern warning coming from God himself would be a deterrent to the first humans. Well, God would find out soon enough, for humans made the mother of all bad choices. Their action was aided and abetted by Satan (the serpent) through deception, but that did not prevent them from facing the consequences of their decision to disobey God. As stated earlier, their disobedience upended God's original plan for creation, necessitating that He implement plan B, if you will. But the fact is that God was not surprised by the action of Adam and Eve. After all, He is almighty, omnipotent, and omniscient. So, why did He not intervene to prevent that epic disaster from occurring? I can't answer that for Him. It shows how crucial free will or choice is to God. Knowing what would happen, His plan B was a redemptive plan to restore humans back to Himself through the work of Christ.

Can you imagine a community where everyone is free to exercise free will, unfettered, whereby people do whatever they like, however they want, and whenever they like? Because we were created to be social and reasoning beings, there is room for us to come together if we so desire, to reason together and achieve consensus in decision-making for the common good. In our modern society, we accomplish this through systems of governance that operate at various levels local, regional, national, and global. For selected areas or aspects of our lives, we agree to a standard code of conduct and craft and enact laws that are binding on all citizenry. We prescribe appropriate consequences for disobeying any law. In a democracy, individuals are accorded certain rights under the law that must be available to all without prejudice or bias. Similarly, the law courts are supposed to be no respecter of persons but treat the poor and the rich the same way. If you are a Christian, it behooves you, as much as it is within your power, to ensure that your neighbor receives fair treatment in society unconditionally.

Note that laws in society apply to only a tiny fraction of our daily decisions. The majority of decisions we make in life are unlegislated. Your neighbor might make a choice that you disagree with. Still, it is their choice nonetheless, and as long as it is not in contravention with the law or common sense social etiquette, you have to respect it and Love them unconditionally. It is alright to offer to counsel them, help them see and make better choices, and pray for them. But you may not force them to make your preferred choice. If God, knowing the grave consequences of the disobedience of Adam and Eve, allowed them to make their own choice anyway, you surely can do likewise. One weakness that has been costly and has frustrated the work assigned to us to go into all the world and invite all people to come to Christ, is our misunderstanding of choice in the life of humans. The consequence is that we drive away the very people whom we think need to for sake their sinful ways by the unjustifiable and unchristian roadblocks we place in their way to salvation in Christ.

I wish to directly address the practice of some in the Church whereby they select a few "sins"as the cardinal sins, as I once heard a pastor say after a lengthy rant against abortion and homosexuality. The popular scriptures to support this position include 1Cor. 6:9-10, where it states that neither fornicators, nor idolaters, nor adulterers, nor effeminate, nor abusers of themselves with humankind, nor thieves, nor covetous, nor drunkards, nor revilers, nor extortioners, shall inherit the kingdom of God." What about the sins that God hates (Proverbs 6:16) haughty eyes, a lying tongue, hands that kill the innocent, a heart that plots evil, feet that race to do wrong, a false witness who pours out lies, and a person who sows discord in the family? Well, please look at these lists (there are others) and see if anyone can inherit God's kingdom as is. Let me be clear if God says (the Bible) it is wrong, it is wrong. Period. No trivializing or excusing it. Christ ushered in a superior New Covenant with higher standards in God's plan B.

He declared, for example, that you are already guilty of adultery if you look at a woman lust fully. On, murder, He said that anyone who is even angry with his brother or sister would come under judgment like the murderer (Matthew 5:21-22).

Bottom line, ALL have sinned and fallen short of God's glory (Romans 3:23) and need a Savior. Just like the case of the adulterer who was caught red handed in the act and was about to be lynched by religious zealots, Jesus intervened by saying, "Let anyone who is without sin among you cast the first stone (John 8:7).There is no vigilantism in Christ. Just unconditional love, grace, mercy and understanding. But for the grace of God, you could be doing the same or even worse things than what you are pointing to about another person.

The good news is that, in Christ, ALL is forgiven; the past, present, and future sins are forgiven! If you confess your sins, no sin is too small or too big to be forgiven. All are welcome. There is no need or justification whatsoever for anyone to try to exclude another person from accessing this free salvation by selecting a sin or two and using them to harass or browbeat another for any reason. At any point in time, each one of us would be struggling with one or more sins. How dare you or I point at or look down upon another person with disdain for the sin they struggle with when we have our struggles? (Matthew 7:3-5). Remember, our role in God's plan B is to bring people to Christ, no questions asked. He saves, not we - the Holy Spirit convicts people of sin, not us. In Christ, we are a new creation. Old things are passed away. In Christ, we begin our journey of transformation into His image. The change of heart comes from God, but we can help each other with the process through prayer, compassion, love, patience, and understanding towards others, but NOT by shaming, hating, or insulting people.

As we confront strongholds of sin in society, Let us not forget that, as Christians, our weapons are not carnal but spiritual (2 Car. 10:4).

Our charge to evangelize the world is the main event, the big picture in God's plan B Led by Christ. Let us not get sidetracked by not seeing the forest for the trees and resort to human efforts (the arm of flesh approach) to reach people for Christ.

The devil is very cunning, and just like his deception derailed God's plan A, he is up to his game again in plan B. He wants to deceive and sidetrack us into following traditions and fighting culture wars instead of waging war for people's souls. We have seen this movie before, The Moral Majority. Released in 1979, it was a spectacular flop at the box office, and rightly so. The agenda was to advance conservative positions on social issues.

We don't seem to have learned any lesson from the abject failure of such arm of flesh combat against spiritual forces. Instead, we have released the sequel under the title "Love the sinner, hate the sin," whereby, through what I call "sin profiling," some people are targeted for what is tantamount to "religious hate crime. "In hating the sin, the sinner feels hated because of the ill-advised tactics. When that happens, we deny people the opportunity to have a genuine encounter with Jesus and become saved and receive the power to overcome their sins. Next time you go to church, please look to your right and then to your left, and you will find a sinner saved by grace which is undergoing transformation because they have some sinful habits to work on. You may disapprove of their choices, but if God was willing to allow Adam and Eve to make the mother of all bad decisions, so can you. I said in a previous chapter that the church is not a haven for people to practice sin, for the scripture says that if we continue in sin, we should not expect grace to continue to abound (Romans 6:1). Our duty is to love all people unconditionally and help others, if possible, overcome their short comings without a condescending attitude. For by the grace of God, you could be in a far worse shape. Let God take His time to deal with each individual in the transformation process according to His schedule.

In Chapter One, I stated that the Bible is God's unchanging truth, unlike a textbook that needs updating from time to time. Its message is timeless. But the fact is that the world that needs this truth is changing at breakneck speed. Consequently, the methods we use to reach people for Christ must be periodically reviewed to be effective. Social science research can help us better understand the dynamics in society and thereby make informed decisions about how to engage our world for Christ. We have to get rid of traditions that are not only unbiblical but unhelpful to our cause and the charge Christ gave us. Let us move past elementary concerns like should a Christian smoke? Drink? Go to the movies? Dance? Listen to secular music? Wear ripped jeans? For some Christians, unless Jesus is specifically mentioned in a song or movie, it is unwholesome to watch. What is wrong with drinking a glass of wine now and then? It is different from being a drunkard. Some may argue that one glass today and soon you will be hooked on alcohol. This may be true for some, so they abstain from alcohol, which is commendable. But don't judge another person and make them feel guilty for drinking a glass of wine, e.g., at dinner.

It is like saying that because God hates gluttony, I will not eat lest I be tempted to overeat and become a glutton. Moderation is the universal caution in such matters. To permit themselves to dance or indulge in secular social activities, some create "acceptable versions" of such actions by attaching the label "Christian" to things - Christian yoga. So, the fashion these days is ripped jeans, tattoos, and colored hair, so what? Are they sinful? No. God looks on the heart and not the outside. Why do you torture yourself with such unhelpful traditions, if I may ask? Frankly, I will not be caught dead in a pair of ripped jeans because I am more traditional in my choice of styles in clothing. But if you like them, knock yourself out. After all, these fashions are so fleeting some last for just one season.

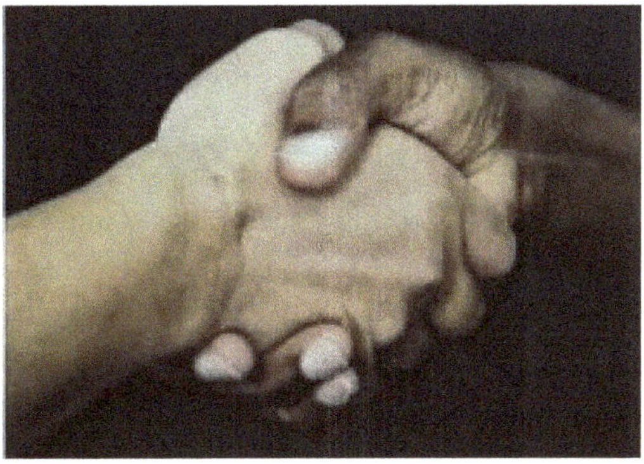

We are the salt of the earth, and to be effective, we must be in contact with the earth. When Jesus was accused of mixing with sinners, he gave this classic response: "Healthy people don't need a doctor but sick people do. I have not come to call those who are righteous, but those who are sinners" (Mark 2:17). We must try to meet people where they are as much as possible. It does not mean you must frequent the strip clubs and attend the shows to reach people who like that type of entertainment. You might get hooked, for sex is a potent attraction. That is why the Bible says the way to handle sexual immorality is to flee. Jesus gave us many examples of choices to make in society, including the few that I have shared so that we can be influential followers of His. Not long ago, the excesses of Hollywood were a popular topic for a Sunday sermon in many churches. Hollywood embodied all that is evil in society. Now, the excesses of some preachers can put Hollywood to shame. "We cannot let the devil have all the good things" is the lame excuse some make when they want to indulge in things they once criticized. If we can undergo such a transformation, we can indeed find ways to accommodate some of the societal changes so that we do not needlessly frustrate our efforts to witness people.

We make STEM based choices routinely and on a daily basis. Credit cards used to be railed against by some preachers as the devil's financial transaction medium. This can very easily lead one on the path to debt if not correctly handled. In this regard, there is every reason to advise people to be very careful about that technology. Similarly, another potentially addictive pastime one must be cautious about is TV viewing. But these days, some stores do not even accept cash for purchases. Many churches now encourage giving via credit cards and other electronic systems like "giving kiosks." We have a popular category of preachers called “Televangelists”. Most churches have online programming that is available 24/7, with the potential to reach millions of people around the globe. They encourage TV viewing and giving via electronic formats, including credit cards. My point is that the influence of STEM in society is only going to continue to increase. Instead of being late adopters of innovation, Christians should be actively engaged in STEM so we can influence its development and the deployment of resulting policies that protect society from fraud or abuse. Technology per se is not evil; people can use it for evil.

Back to morality and ethics. Morality refers to standards by which society judges what right or wrong or good or bad behavior is. These standards are dictated by society's culture, religion, or philosophies. The standards enable people to live and cooperate with community members, sometimes sacrificing self-interest for the common good. In practice, one tends to act based on personal beliefs of what is right or wrong, even if they are different from society's views. Consequently, morality is more of a personal principle of conduct.

We often hear the sentiment "You cannot legislate morality" expressed in society. It is an affair of the heart (individual principle). Christianity is about changing hearts first, so good conduct and behavior will manifest in one's life.

On the other hand, ethics may be defined as standardized codes of behavior prescribed by asocial system (external principle) to obligate people to conform to a higher standard of conduct (what ought to be versus what is). Ethics demands that scientists be honest in their practice. "They are expected to report authentic results rather than withholding relevant information" (Hassan Zohoor). Promoting science and the growth of moral values is necessary for human development. We sometimes hear of the results of industry sponsored research being called into question.

The implication is that because the researchers were paid, there is some pressure to deliver favorable results to the sponsor. There are standards like "conflict of interest, "which urge people to stay away from conduct that could have the appearance of evil. To protect the integrity of the scientific enterprise, researchers are required to disclose any previous activities that could cast doubt on their work. As part of the moral obligation ascribed to scientists, it is expected that they would treat their living subjects, both human and animal, humanely. I discuss this subject further in Chapter Five.

"Lusting after fame or recognition, egoism, greed, prejudice, snobbishness, racism, and political considerations have frequently resulted in immorality in the domain of science" (Hassan Zohoor). Scientific findings can lead to positive or adverse applications that can impact life on Earth (see Chapter 14). For example, some discoveries have polluted the environment and endangered humans and animals.

To their credit, many scientists exercise their ethical values through researching and discovering ways to protect life on Earth and foster healthy societies. Ultimately, society is better served when STEM, the engine that drives our modern economy, incorporates humanitarian and ethical values into its domain that does not jeopardize or compromise its intrinsic values. A cautionary note to all scientists: Just because it can be done does not mean we should do it. We serve society and must be sensitive to the concerns of people. For example, at this time, genetic manipulation of the germline (e.g., to create designer babies) is considered out of bounds to researchers and cannot be supported with public funds.

Combined, morality and ethics would allow a society to operate under norms and values that alert people to right or wrong conduct for the benefit of society. Science is a human endeavor and hence is impacted by human values. Scientists are urged to integrate scientific values with other ethical values honored by society.

The consequences of science and the process of discovery or application of the field are subject to an ethics review. However, some scientists argue that science is morally neutral, concerned with questions of fact and theory and not meaning and morality, and should therefore not be subject to moral judgment.

There must be an untruth if there is truth, otherwise commonly called lies. Society grapples with the concept of absolute versus relative truths. Christians believe that God's truth is absolute. However, others believe that truth is relative and should be personal, each one determining it based on how they feel or according to their belief systems.

Some politicians have been accused of suggesting such a thing as an alternative truth. It is the role of Christians to make God's truth known to the world and persuade people to receive Christ as their Lord and Savior. It is not correct or fair for us to expect unbelievers to conform to our living standards until then. If a dog barks, it is standard and expected behavior. However, if a cow barks, then we have a problem. The onus is on Christians, who purport to know the only truth, to act or live accordingly; otherwise, we cannot expect to exert any credible influence on society. In America, some tend to want to impose Christian standards on society. This stems from the historical fact that America was founded by people fleeing religious persecution in Europe. But the bigger truth is that numerous other immigrants had since joined the party, not to talk about the Native Americans here before all of us came over. Each group came with its customs and beliefs. As a Christian, I believe and live by God's standards, the highest of which is to love God and all people. Both religious and political truths are choices people make. However, natural truths (scientific) are imposed upon society (no respecter of persons). We have to understand STEM to know God's infallible truths revealed via the scientific query of His "Book of Nature."

CHAPTER 13

WHEN STEM GOES WRONG

I discussed how practitioners of faith and politics could go astray in their decisions and actions in previous chapters. Well, science has its fair share of mishaps. The creator or inventor is the one who ascribes the original purpose to an invention. When I talk of STEM gone wrong, I don't mean the willful abuse of an invention to commit evil acts. Great inventions are routinely abused worldwide by people with evil intentions. Even good people do make mistakes. By nature of the field, scientific truths are discovered stepwise, whereby what is true at one time may be replaced when new information becomes available. It should be understood that all new inventions will go through a series of refinements to be perfect. Failure is part of the path to success. During the early part of the COVID-19 pandemic, the scientific community was eager to find the source of the virus so that appropriate actions could be taken to contain it. The initial allegation was that a Chinese research lab was the culprit that may have accidentally released it. If that is true, that would be an excellent example of how scientific research, intending to solve problems, creates one.

One of the most transformative innovations or inventions in recent history is the Internet. This computer-based invention is a globally connected network system that uses a set of networking protocols (called Transmission Control Protocol/Internet Protocol-TCP/IP) to allow two or more computers to communicate via guided, wireless and fiber optic technologies. The players include private, public, business, academic, and government entities that help coordinate and implement standards, protect users, and create innovations to keep the network functioning properly. Originally designed to facilitate the exchange of raw data among researchers, it is now heavily utilized by the so called social media platforms (Meta, Twitter, etc.). One of the universal cautions about the Internet is, "Don't believe everything you find on the Internet." This is because the revolutionary invention is open to both good and bad actors. One pastor lamented that the Internet provides ready access to the Bible as well as pornography. There are what we call "cybercrimes, "vices committed on the Internet. Cottage industries keep springing up, designed to defraud users. Unscrupulous politicians can abuse social media to cheat and steal elections (e.g. spreading dis and misinformation). Foreign influence in political elections is now a reality. Terrorists and hate groups use the network for diabolical activities. Many a youth have suffered from cyberbullying. Individuals, governments, and corporations have suffered at the hands of malicious electronic viral attacks launched by individuals or sometimes even state sponsored groups for espionage or economic sabotage. In this particular example, the inventors of Internet technology meant for it to facilitate good communication. Still, the unintended consequence is providing the same opportunity for bad actors to operate for evil outcomes.

The discovery of fossil fuels (oil, coal, gas) provided energy to drive the global industrial revolution. Petroleum and petroleum-based products are pervasive in modern society, ranging from their application in cars to cosmetics. But the by products from their manufacturing and applications have adversely impacted the earth's climate. Pollution from the burning of fossil fuels (carbon monoxide, nitrogen oxides, sulfur oxides, and other hydrocarbons) is a health hazard and a source of economic loss to the nation. Fossil fuels contribute to the so called greenhouse gases (carbon dioxide, carbon monoxide, methane, and fluorinated gases), which are implicated in the current global challenge called "climate change," the change in the usual weather pattern found in a place (could be regionally or globally).

Climate change affects the melting of ice, resulting in rising sea levels, which threatens coastal habitats. Temperature changes are affecting wildlife and their habitats. Diseases are appearing in places where they were not common before climate change. The problem is blamed for the increasing frequency and severity of wildfires that plague places like California. Climate change is best tackled as a global challenge because the wind blows wherever it may. Nations that pollute the least tend to suffer the most because they lack the technologies to combat the adverse effects of pollution. Unfortunately, politicians tend to make decisions based on selfish interests, not neighborliness.

One invention that changed the world for good was plastics. Plastics are pervasive in modern society and have been of tremendous benefit in the packaging and containerizing of products. They are found in cars, toys, furniture, etc. Unfortunately, one of the biggest challenges in global pollution is the disposal of plastic waste. Rivers and other water bodies are being choked by plastic containers that are not properly disposed of. When even disposed of at trash sites, the bigger problem is that plastics are not susceptible to decomposition by microbes, as other organic materials are, and therefore cannot participate in existing natural nutrient cycles. After decades of using plastics, we woke up one day to learn that under certain conditions, some toxic chemicals in the material can leach into the food or drink that plastic containers hold.

Life expectancy has been significantly extended for humans due to technological innovations in medicine and the health industry. Drugs are available for curing various diseases and alleviating pain. Unfortunately, while of tremendous use in medicine, some drugs have severe adverse consequences that have devastated lives in society. Some users of certain prescription drugs have become addicted to them. Sadly, the use of particular drugs has been deliberately perverted by unsavory individuals. Illegal access and improper use of prescription drugs have resulted in deaths from overdose. Some manufacturers and healthcare providers have resorted to profiteering from life-saving drugs.

One invention that has been perverted in society is guns. Available in a mind boggling variety, guns are suitable for hunting, warfare, and recreational purposes. Unfortunately, some of the saddest events in society are gun-related. Some politicians and gun lobbyists collude to make guns readily accessible to all who qualify to own one. In the hands of criminals, they are used to commit offenses against society, such as armed robbery and the killing of police. Unfortunately, in the hands of law enforcement, some have abused them to commit tragic killings. At home, some are quick to resolve domestic issues with a gun. Assault weapons belong on the battlefield, not in the community. We often hear about the need for "common sense gun laws." Still, lawmakers often lack the courage to go beyond acknowledging the problem and formulating new and effective laws or enforcing existing ones. The voting public is partly to blame for the state of affairs.

Microbial-based biotechnologies have been tremendously beneficial to society, medicine and health, food and agriculture, and industrial settings. But these technologies have been used for warfare, causing the death of untold millions. Biological and biochemical weapons are notorious for the inhumane manner in which they cause death in their victims. Thank God the world no longer accepts these weapons as appropriate for conventional warfare. Unfortunately, unscrupulous political leaders have been known to use them for terrorism in modern times.

In 1970, a devastating disease epidemic occurred in the corn industry. The southern corn leaf blight, a fungal disease, was a menace in corn production. As previously discussed, one of the breeding techniques responsible for boosting crop productivity is the hybrid seed. According to the USDA, in the 1930s, corn yields were about 22 bushels/acre, obtained from using conventional varieties. The switch to hybrid seed increased corn yields to nearly 200 bushels per acre in the 1970s. But as previously explained, the production of hybrid seeds is very tedious. Initially, one plant had to be manually detasseled (the male, tassel, had to be removed to make the plant female) to allow wind pollination to be used in the field. Researchers eliminated the need for detasseling by incorporating a genetic male sterility mechanism (Cytoplasmic Male Sterility, CSM-T) into one parent (the tassel, male flowers, were rendered sterile genetically).

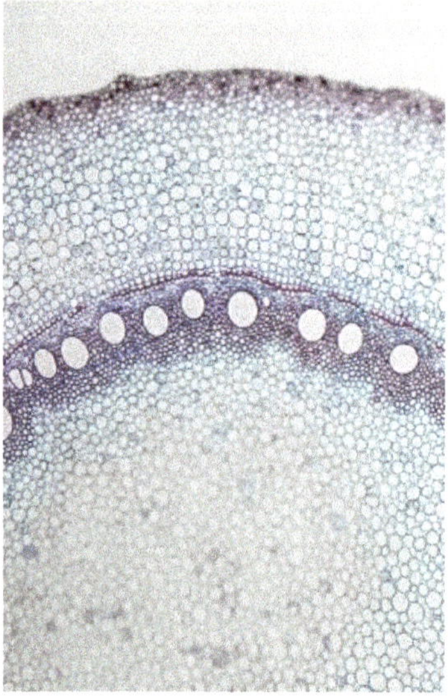

Most corn producers adopted the hybrid seed for high productivity.

Unfortunately, in 1969, a new, unfamiliar corn leaf disease emerged in production. The hybrids using the CSM-T system were disproportionately susceptible to this new disease. Consequently, the hybrid industry was devastated, leading producers to hasten back to the old method of tedious detasseling to create hybrids until more improved techniques emerged.

Modern history is punctuated by numerous engineering fails that have resulted in the loss of numerous lives. Engineers thought they had made the correct calculations in their designs of bridges, buildings, dams, etc., but they were rudely awakened by disasters that ensued because of human errors or unanticipated problems. Sometimes, engineers cut corners to save money or they did not invest sufficiently in feasibility studies. Many of the spectacular achievements in STEM have come by literally defying the Laws of nature. To harness the power of water for hydroelectricity, engineers artificially dam rivers with various consequences. Some homes and lands upstream can be flooded. But in the event of a breach of the dam, untold damage to life and property has been known to occur downstream. A similar situation occurs with nuclear power accidents. Whereas nuclear technology has been successfully applied to generate power for domestic use, nuclear accidents, when they occur, have effects that linger for decades.

Specific examples of engineering failures include the following. A) The 1937 Hindenburg disaster that claimed 37 lives was caused by an electrostatic charge that ignited the leaking hydrogen gas. B) The unsinkable Titanic sunk upon colliding with an iceberg because more than the touted survivable breaches of four compartments occurred.

It was also speculated that low-quality steel or weak rivets contributed to the flooding of the breached compartments. C) The deck of the Tacoma Narrows Bridge moved vertically under windy conditions after opening to traffic, and eventually collapsed in 1940. D) Structural failures were to blame for the Space Shuttle Challenger disaster in 1986.

The accident was blamed on the resiliency of the O-rings following a severe cold event that swept through Florida the night before the launch. E) In 1981, the two walkways of the Hyatt Regency Hotel collapsed and killed 144 people. It was blamed on design flaws in the steel hanger rods supporting the walkways. F) The Charles De Gaulle airport terminal collapse in 2004 was blamed on inadequate feasibility studies that contributed to design flaws. G) The failure of the Banqiao dam in China in 1975 was blamed on poor engineering design and subsequent poor maintenance. H) The drug thalidomide created in Germany was widely prescribed to pregnant mothers to fight morning sickness until it was linked to physical deformities in infants. It was banned in 1961. I) The Walkie Talkie Center in London, UK, is an award-winning building with serious environmental consequences. The much-celebrated concave design is blamed for concentrating sunlight onto some streets and posing a health hazard. The defect was corrected by fitting it within sun shades. J) Mathematicians are not immune to errors. Sometimes, it can be as simple as misplacing a period, which happened in Spain's 2003 S-80 submarine program. After one of the subs had been nearly completed in 2013, it was discovered that it was 70 tons too heavy and in danger of being unable to resurface after launch. The error was blamed on a misplaced decimal point that was not discovered during the design stage. K) Another common math error occurs during data conversion between international measurement systems. In 1983, Air Canada Boeing 767 flight from Ottawa to Edmonton crash-landed because of a loss of engine power from insufficient fuel. Ground crew in Canada fueled the plane based on the imperial system, while the Boeing aircraft, made in America, was designed on the metric system. L) Sometimes, humans blame the computer for mathematical errors. In 1996, the European Space Agency's Ariane 5 rocket exploded soon after launch because of what was called an integer overflow error in the software used for the launch. M) NASA has had its issues with data conversion attributed to the possible crash of the Mars Climate Orbiter.

Lockheed Martin's programming software used the imperial measurement system, while NASA used the metric system. This list provides a gist of the mistakes that occur in developing and applying STEM in society. These mistakes occur worldwide irrespective of persons, nations, or the stature of corporations.

Scientists are human and can make mistakes. Unlike God, who is all knowing, there is a lot yet to know and understand about nature, the blueprint for science. Deciphering the book of nature is a work in progress. Applications of scientific knowledge involves trial and error until perfection is achieved. To protect the citizenry from errors that scientists may make, the government and professional bodies have instituted standards as guidelines in manufacturing, construction, and product development that practitioners must adhere to. These are based on current scientific information and are periodically reviewed and revised as new knowledge becomes available.

Unfortunately, sometimes, we learn the hard way in the classroom of nature and are humbled by an F grade, when our technological and engineering marvels prove to be no match for the forces of nature. Then, we are compelled to go back to the drawing board and impose higher standards on our operations. Standards notwithstanding, some humans can deliberately circumvent guidelines for various reasons including profit. Bad actors can abuse knowledge and products from science technology and engineering, for destructive purposes. Defective products may be recalled, but sometimes only after the loss of lives and property damage has already occurred. I will close this chapter with a few proverbs from my native Ghana. Shall we deny ourselves sleep, because we might die in our sleep? No, we shall not. Shall we deny ourselves a smoke of tobacco, because we are carrying gunpowder? No, we shall not. So, to the whole world, shall we deny ourselves the benefits of science, technology, engineering, and mathematics (STEM) because of possible mistakes by scientists and deliberate evil acts by bad actors? No, we shall not. LET US THANK GOD FOR SCIENCE.

CHAPTER 14

LIVING BY FAITH, SCIENCE, AND POLITICS IN SOCIETY

In previous chapters, I identified three key forces that influence our lives in society faith, science, and politics, and discussed how each works. We followed it with pairwise comparisons of these domains. This chapter will consider how all three interact in society where the rubber meets the road, if you will. When all is said and done, my goal is for all, especially Christians, to understand that God created nature, and science is the way to understand it. As we understand how nature works, we nudge it to our advantage. This, I believe, is God's plan that He embedded in creation. The evidence is incontrovertible that STEM is God's way for us to have dominion of the earth.

To start the discussion, let me point out that a person can engage the three domains of faith, science, and politics at one time. The challenge then becomes how one balances the unique operational guidelines and principles that pertain to each domain that we reviewed, to be successful in life. As stated in my introduction to this book, the goal of this book is to encourage people of faith, especially Christians, to embrace STEM (Science, Technology, Engineering, and Mathematics) as part of God's plan for His creation. This way, they may see that God can use science for His purposes. Before we delve into how these entities interact, we need to review certain basic characteristics of society and how it operates.

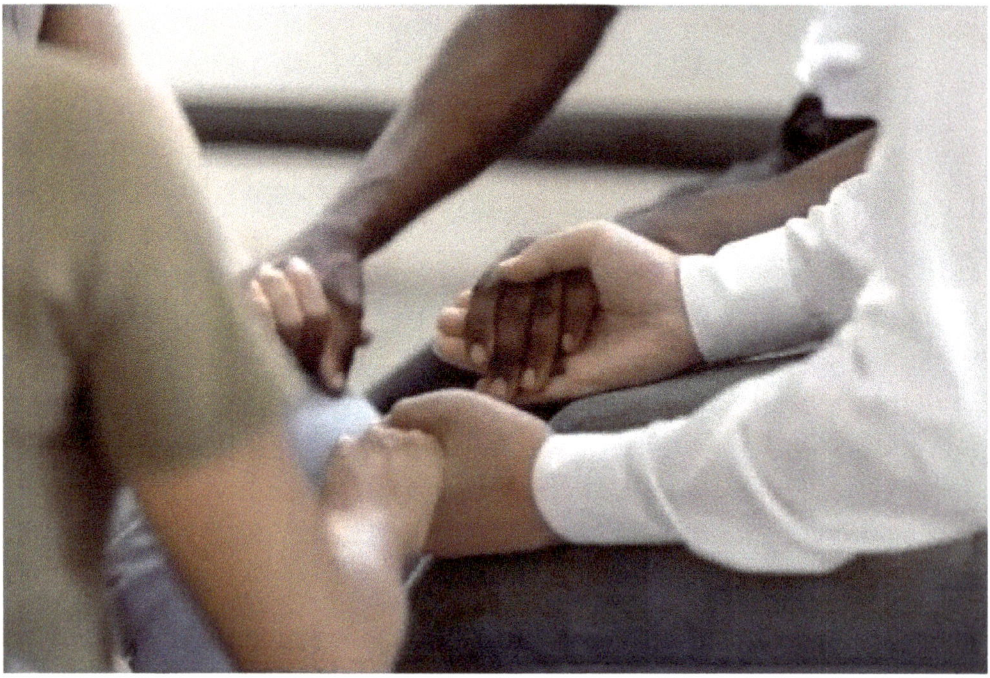

To briefly recap the information from the previous chapters, scientists voluntarily work in groups that are governed by policies and guidelines. They have the authority to discipline their members when they go astray. The method of science is the universal guideline for how scientists conduct their affairs. No single person pontificates over or administers this protocol for the global scientific community. Numerous professional associations operate under their specific guidelines to ensure that their members maintain high integrity in their operations. Each individual, at any time, may produce evidence to challenge any information put out in the public place. The preponderance of the evidence is the arbiter of scientific truth. In Christianity, the Bible is the final authority on truth. Christians also gather in groups, originally called churches in the Bible.

In our modern society, churches have clustered under denominations, each with its interpretation of Scripture and administered at various levels by leaders who preside over them and can exercise disciplinary actions against members. Science and Christianity exist in a society that operate sunder a political system with its own truth, if you will. Political truth is the law of the land. The question then is, whose truth must be obeyed in society?

There is no one to tell you to obey natural laws. The fact is natural laws are no respecter of persons. On the other hand, religious and political authorities are susceptible to social influence and manipulation. In my opinion, the role of religion in society is to change people's hearts and thereby help provide a moral compass to guide how politics and science operate for the good of all. At the end of the day, people in society submit to the ruling political authority. In a democracy like the US, all citizens are subject to the laws of the land. But what do you do, as an individual, when the laws of the land contradict the laws of your religion or profession? What do you do as a scientist when politicians disregard scientific facts in their decision making? I don't have any earth shaking answers for these situations, just some suggestions. For the Christian, your first step is to ask yourself, "What Would Jesus Do? (WWJD). There are principles and concepts in Scripture to guide our behavior in all situations. Jesus gave us some examples that we would do well to follow.

In dealing with politics, Jesus said this on one occasion: "Render unto Caesar that which is Caesar's" (Matthew 22:21). Simply put, we are all expected to respect state authority and, in the specific context of the quote, to pay our taxes. Another pertinent scripture states that: "Obey the government, for God is the one who has put it there.

There is no government anywhere that God has not placed in power "(Romans 13:1-2)." Further, there are provisions in the constitution for individuals or groups to protest the action of politicians and even seek redress through the courts. There is such a thing as the ballot box, which terrifies politicians.

Voters can exercise their rights by voting out politicians who fail to perform to their satisfaction. For the Christian, our actions should be consistent, regardless of the political party in power, and be guided by Scripture.

As individual human beings,we are unique in our genetic makeup, as science has revealed. Each one is one of a kind. This uniqueness makes us crave to be independent in our ways in life. More importantly, God gave us free will or the capacity to choose between different courses of action unimpeded. Being communal beings, we tend to live in groups. To do so successfully, groups have rules or laws that govern the conduct of members. While we may maintain and insist on some personal freedoms, we must sacrifice others to live together as a group. For a nation to progress and thrive, we need competent political leaders committed to serving the people and not themselves. Then we need religious leaders who can speak truth to power, stand up for the truth (all truth, not selective) all the time, and love all people (unconditionally). And finally, we also need scientists who will work in the service of humanity, respecting the moral standards of society. Knowing the pervasiveness of STEM in society, we must elect political leaders who have some science education, or at least have appreciation and respect for science, and are willing to listen to professionals.

Leadership is everything. The attitude of a society or nation reflects its leadership. Because of the nature of politics, a nation where religion and politics are bedfellows can be a recipe for disaster. There is nothing Christian about partisan politics. Christianity is rooted in selflessness, while politics is rooted in selfishness. When politics and Christianity get into bed, the union is unequally yoked. This ungodly compromise often leads to the voice of truth being silenced or sidelined. All these observations were on display during the Trump administration when a segment of Christians (White Evangelicals) declared their unflinching support for him. Consequently, some leaders kept quiet when the voice of truth needed to be heard.

Being a Christian majority country, it is easy to see why people will vote for politicians who promise to uphold morality in society. Morality, the code by which the community determines what is right or wrong, should feature in the decisions made by lawmakers. Can anything be more potent than belief or faith? It is not based on objective truth. How do you dissuade someone from taking a course of action that he or she believes was mandated by God? Scientists are ill equipped to change the hearts of people. Unfortunately, morality is usually defined in narrow religious terms and hijacked by some to mean a few selected areas of our social life (sexual morality, sanctity of life). "True religion or the religion that God accepts as pure and faultless is this: to look after orphans and widows in their distress and to keep oneself from being polluted by the world" (James 1:27). Many cross-references in the Bible dearly emphasize that society should pay attention to the plight of those who are marginalized or disadvantaged among us.

Politicians, who want to be truly moral, should make decisions that eliminate distress in our nation (e.g., poverty, injustice, discrimination of all kinds)."The world does not need a definition of religion as much as it needs a demonstration" (Martin Luther). Faith without works is dead (useless).

Before I go any further, let me clarify that it is not my intention to paint politics with a broad paint brush, declare it an evil institution, or present all politicians as unscrupulous. Absolutely not! There are decent men and women in politics who are courageous and willing to break ranks with colleagues in their party and put the country first. Yes, Christians can and should participate in the political process at all levels and be guided by godly principles (not religious dictates), not blind allegiance to a political party. Wherever you serve as a Christian in society, you are Christ's representative, salt of the earth, a change agent for good. WWJD? Many Christian leaders are honest and faithful to their calling. They are true followers of Christ and not religious traditions that nullify the gospel (Mark 7:13). However, some are best described as misleaders because they misinterpret scriptures and their willful politicization, commercialization, and socialization of the faith.

Now, I will discuss specific examples of how science, religion, and politics intersect in society. The Declaration of Independence, which was published in 1776, contains a powerful statement that is very popular in society: "We hold these truths to be self evident, that all men are created equal, that they are endowed by their Creator with certain unalienable Rights, that among these are Life, Liberty and the pursuit of Happiness." The First Amendment allows US citizens the freedom to practice any religion of their choice and prevents the government from officially recognizing or favoring any religion.

In essence, the government must maintain an attitude of neutrality towards religion. The statement I just read from the US Constitution places Christianity front and center in politics and governance, not necessarily as direct participants, but at least, as Christ did, to advocate for equality for all men and women, and especially justice for the poor and those who are socially disadvantaged. Whereas the government must not favor a particular religion, political parties, according to their nature, are under no obligation to do the same. The GOP openly touts Evangelicals (especially White Evangelicals) in the US as its political base. In return for its support, the GOP favors this base with the appointment of the so called conservative judges, supports an abortion ban, and is unfavorably disposed to gay rights. I believe the only person who belongs on the Supreme Court, and for that matter, any other court of justice, is one who will administer justice and fairness to all and not play politics from the bench. As the Bible teaches, a judge should not be partial to the poor or the rich.

As individuals, the Bible strongly cautions against judging others: "Judge not, lest you be judged" (Matthew 7:1)."

"Why do you worry about the speck in your friend's eye when you have a log in your own?"(Matthew 7:3). The Bible calls this hypocrisy. I don't think the Bible forbids all manner of judging. What it cautions against is judgmentalism. It is alright to be critical by examining an idea, statement, or behavior in a well-reasoned fashion. Being judgmental or criticizing is taking a position on an issue based on your personal value system. Do not judge others by your own standards, lest you treat people with a "holier than thou attitude," which could degenerate into hate.

In a democracy, Christians may participate in any aspect of our society in which they have an interest politics, government, education, medicine, scientific research, police work, entertainment, sports, etc . But as previously stated, wherever we find ourselves, we are expected to be Christ like in our conduct, which boils down to loving God and loving people, all people. We don't have to agree with people on all or even any issue. But because they are human beings, we are commanded by God to love them unconditionally, as Christ does, and not be judgmental. In politics, we can be Democrat or Republican in our affiliation, but in the end, it should all be about Christ, not liberalism or conservatism. Without Christ, what do we liberate? Our carnal liberties shall consume us. Similarly, without Christ, what do we conserve? Garbage. Conservatism does not equate to Christianity, as some make it sound. When we are partisan in our politics, we face the danger of losing our moral leadership in society. There is the temptation to be silent for fear of losing political clout. Let us be reminded that politics was not Jesus's cup of tea anyway. The people at the fringes of society, those who had been marginalized and overlooked by religious and secular institutions, were His business.

There is no denying that in the US, only a handful of Christian leaders have access to major communication platforms, such as radio and TV, and hence have access to the ear of more of the public on matters of social concern and significance. Needless to say, many of these "mega leaders" with megaphones often espouse the so called conservative views on social issues. But as I have previously indicated, if you mount the pulpit in the name of Christ or purport to present Christian views, it can only be about the truth, not partisan political talking points. Two major events of 2020 would help illustrate my point. While in the grips of the COVID-19 global pandemic and the US reeling under its devastation, some politicians resorted to playing partisan politics with that significant health event. They were oblivious that the virus was no respecter of persons.

Instead of persuading their audience or followers to listen to scientists, the silence of these leaders was deafening. I saw or heard only a few mega Christian leaders who were bold enough to challenge Christians to at least use common sense and respect the advice given by the knowledgeable scientists and professionals, not politicians, on staying home until it was safe to congregate. This sad response or lack thereof to such a devastating event was due to the cozy alliance of some Christian groups with politicians, which paralyzed these leaders and prevented them from speaking truth to power.

I guess some Christians crave political power rather than following Christ, our Master. What was gained in return for their silence? Nothing of significance to the cause of the Gospel but embarrassing exploitation of Christianity by politicians for selfish gain.

Then came the killing of George Floyd by the police, which exposed the systemic racism and injustice perpetrated against African Americans. Tons of social science research has shown in equities in American society. There was outrage expressed around the world for weeks in never before witnessed ways. White Americans sometimes dominated these protests in the US. Again, the usual suspects," front and center on pet issues, were nowhere to be found. An opportunity was missed, yet again, for Christians to unite to show the world what it means to be a true follower of Christ Love God, Love people, Love justice, Love truth, Love righteousness, Love grace and mercy, stand up for the poor and those who are socially disadvantaged. Sanctity of life applies to all life, from the cradle to the grave. This is true Pro Life. How can you say you love God whom you don't see, when you don't Love your neighbor whom you see? How can you say you Love the unborn you don't see, when you are blind to the suffering and injustice of those you see? Some of these leaders were quick to come to the defense of a murderer of an abortion provider but kept mum when a policeman killed George Floyd in such a manner that shocked the world's sensibilities. As Christians, we should be careful not to drive away or hinder those we seek to win over into the family of God by counter intuitive actions.

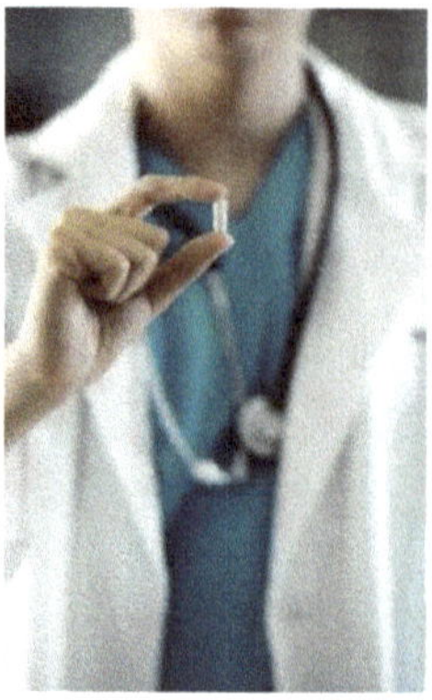

Some in society, including doctors, disagree on abortion, whether it should be performed at all or at what stage in pregnancy. The Pro Life movement generally has a zero tolerance stance on abortion. The position stems from the belief that life begins at conception. Science supports the fact that fertilization (union of sperm and egg) is the first step in the development of human life. Some Christians like me, who do not support abortion as a routine birth control option, believe that there are circumstances under which abortion is justifiable. My wife, an OB/GYN, will not prescribe the so called Morning After Pill, not being certain if fertilization has occurred or not. On medical grounds, when the mothers life is in danger, abortion may be performed, provided it is with the consent of the woman.

Similarly, when a woman has been raped, the choice has to be hers as to whether to keep the baby or not. Whereas I don't support the use of abortion as one of the routine pregnancy prevention approaches, I support the use of scientific contraceptives to prevent conception. Those most vocal about abortion would be more credible if they were at least as vocal about the plight of mothers and children born out of wedlock who languish in poverty. Advocacy for the sanctity of life applies, perhaps more so, to life outside the womb than life that may or may not even be naturally viable. Further, when a woman opts for abortion, for whatever reason, it is not our place to judge her. It is an agonizing decision that could have lasting and sometimes traumatic effects on the woman. We owe her nothing but love and support.

The big message here is that our life is a series of choices that we make, with good or bad consequences. God gave humans the right to choose in their decision making. The choice is the individual's to make. If God was willing to stand back and watch Adam and Eve make the wrong choice that altered the course of humanity for the worse, who are you to deny another person their right to choose? You can counsel, educate, help, and pray for people to make what you believe are good choices. If they don't, you have an obligation to love them just the same. During the COVID-19 pandemic, the nation was thrown into turmoil over vaccine mandates. Declarations such as the following were common on TV: "We are Americans, no one tells us what to do"; "My child, my decision"; My body, my choice." How is that different from someone saying, "my life, my body, and my choice"?

The other abortion related controversy is the stem cell technique. After fertilization, the cell under goes repeated division to produce a mass of cells (unspecialized or undifferentiated). These cells start to specialize (differentiate) and develop into the different tissues of the organism. Stem cells from this origin are embryonic stem cells and are at the heart of the abortion controversy. The body uses these primitive master cells to renew and repair its damaged tissues. Researchers may obtain embryonic products from hospitals or abortion clinics, and retrieve stem cells from them for their work, a practice that is frowned upon by the Pro Life movement, which believes it encourages abortion.

Consequently, researchers have sought stem cells from somatic (body) tissues, called adult stem cells. One of the most successful applications of adult stem cells is in bone marrow transplants. Research is ongoing to determine the applicability of stem cells for treating diseases like diabetes, heart disease, Parkinson's disease, and spinal cord injury.

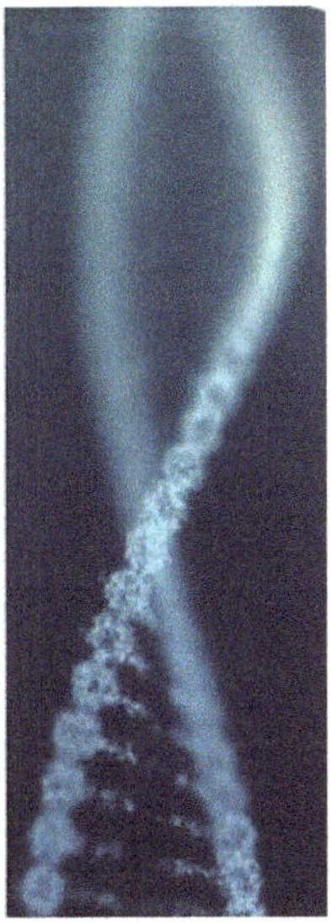

Gene therapy is an experimental technique used to transfer genetic material into the cell to correct abnormal genes that cause diseases. Researchers accomplish this goal by several approaches: replace a mutated gene with a healthy copy of the gene, in activate (knock out) a mutated gene, or introduce a brand new desired gene. They genetically engineer a virus (vector or carrier) to contain the desired gene and allow it to infect the diseased cell (enter the nucleus). The DNA may be integrated into the nucleus only but not inserted into the chromosome, depending on the virus. This technique has potential for treating diseases, including cancer and some viral infections, but unfortunately remains a high risk procedure. Like stem cell research, genetic therapy needs to be supported by the government to proceed safely and ethically.

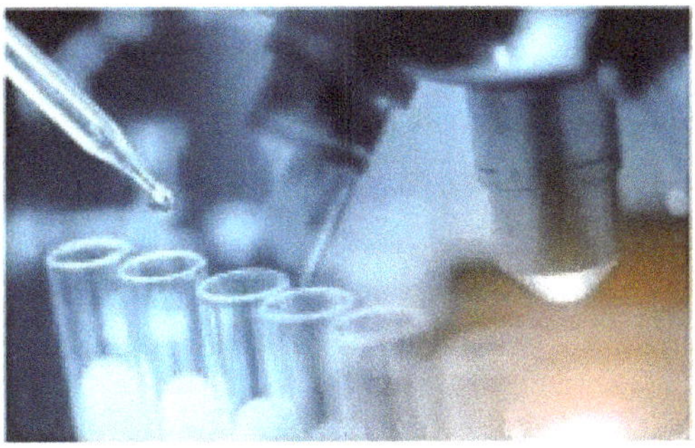

Genome editing (gene editing) is a set of techniques used by scientists to change the DNA of organisms. This is just like editing an MS Word document on a computer, whereby typos can be corrected, words deleted, or new words added. There are several gene editing systems in use in research. One of the more precise and most recent is the CRISPR-Cas9 system (Clustered Regularly Inter spaced Short Palindromic Repeats). Incidentally, this innovation was awarded the 2020 Nobel Prize for Chemistry! This is yet another technique scientists learned from the wild, where some bacteria use this mechanism as natural immunity against reinfection. CRISPR is a group of DNA sequences found in the genomes of some bacteria following a viral attack.

Yes, bacteria can also get sick. These sequences are derived from viruses that infect bacteria (bacteriophages), and are used to detect and destroy bacteriophages during subsequent infections (natural anti viral). The CRISPR-Cas9 system allows scientists to cut out a gene segment and replace it with another. This technique has the potential for addressing diseases including cystic fibrosis, sickle cell anemia, and Huntington's disease. However, it came under severe scrutiny when it was used to edit the human genome, the most controversial being the revelation by a Chinese researcher that he had apparently edited the genomes of twin girls. For all the potential of this technique, I believe editing the germline should be prohibited by the government. The idea of "designer babies" and super-humans is unethical. It is heartening to know that the scientific community does not support such radical applications (editing the germline).

In 2020, the Trump administration instituted Operation Warp Speed as its official unit to guide the development and deployment of a vaccine (or vaccines) to combat the COVID-19 virus. As promised, vaccines were developed, evaluated, and approved in record time. Naturally, the accolades went to politicians who lived for such moments. Scientists should have been lauded for the monumental achievement. But in a way, it was not surprising for the very few people who know how that scientific feat was accomplished in such a short time. The truth is that scientists have worked for decades on the coronavirus and have practiced with epidemics involving this virus in the past. Remember the SARS and MERS epidemics of 2003 and 2012, respectively? Coronaviruses caused these. In fact the COVID-19 is caused by a new coronavirus called SARS CoV-2. Scientists are intimately acquainted with this virus and had platforms for developing vaccines already in place. The Oxford-Astra Zeneca vaccine uses a traditional vaccine platform or technology. A virus (adenovirus, which causes the common cold) is genetically engineered to carry (vector) the gene that causes the characteristic spiked protein (hence the "adenovirus vector vaccine'). The virus is inactivated, so it cannot cause disease (the common cold). But once injected into humans, it tricks the cells into producing the spiked protein. The Pfizer and Moderna vaccines use a modern technology, the mRNA (messenger RNA has instructions for forming proteins) platform (hence "mRNA vaccines"). The synthetic mRNA is injected directly into the human cells (no use of a live virus as a courier). It initiates the production of the viral proteins without entering the cell's nucleus and becoming part of its genetic material. The cell breaks down and gets rid of the mRNA after using the instructions. The Oxford-Astra Zeneca vaccine is cheaper to produce and store (in a common refrigerator) and hence is promising for use in developing countries. The Pfizer and Moderna vaccines are pricier and need extremely cold storage.

The examples of genetic engineering that I have cited here and else where use techniques that exist in nature. But I believe, along with others, that just because we can do it as scientists does not mean we should. The fear of unintended consequences is always a concern when conducting gene engineering. These techniques involve manipulating bacteria and viruses, organisms that evolve rapidly. What if antibiotic resistant, disease causing organisms are inadvertently created? The power of some of these techniques can be abused (e.g., for biological warfare). Whereas I support these technologies, I also believe that scientists must act responsibly to ensure that, in an effort to do good, we do no harm. That said, I believe wanton destruction of research labs and harassing researchers who conduct such research should be strongly discouraged. The general public should rest assured that government agencies monitor such research to ensure the safety of all people.

Climate change is one of the controversial topics in society today. Despite the overwhelming scientific evidence pointing to the significant contribution of human activities to the problem, some politicians deny this fact. In fact, a vivid show and tell was offered to the world during the 2020 COVID-19 pandemic. As a result of the mandatory stay home orders, many vehicles were kept off the roads for a period, reducing the normal presence of pollutants in the air from the burning of fossil fuels. Consequently, places like Beijing and Los Angeles, which generally have very poor visibility due to pollution, showed clear skies and high visibility for all to see for a while. This did little to convince the naysayers of the role of humans in contributing to the pollution that causes climate change.

The so called dirtiest fuel, coal, is so politicized that no intelligent discussion can be had on the need for clean fuels. God instructed humans to take charge of the earth. So, I would expect the Christian community to be a strong advocate for a healthy planet for the sake of humanity. Unfortunately, very little vocal support exists from religious groups or leaders.

There are ways in which church leaders can speak out on social issues without being political or running the risk of losing their tax exempt status. Truth hurts, but it must be spoken at the right time and with love, compassion, mercy, and grace. We know that some politicians will do whatever it takes to get into or remain in power. Some lie habitually and will cheat for political gains. Just like the prophets of old who boldly spoke truth to power in the name of God, leaders of today must do likewise in the face of injustice and other social ills. But when Christian leaders form cozy alliances with politicians for political influence, they lose spiritual capital. During an election season, some pastors turn their pulpits into political platforms. When political pastors speak, it is only against their political opponents. The "sins" of their favorite leaders are pardonable, but they won't hesitate to turn the "sins" of their opponents into political sermons. Partisan politics is beneath the church. It weakens our witness as followers of Christ. Such behavior makes us keep at bay those who need to be brought into the family of God. Remember, God is neither Republican nor Democrat. Coercing people to vote based on human selected single issues is not Christian.

Scientific research produces information that can be acted on for social transformation. Social science research encompasses social and primary care, business, economics, education, population structure, unemployment, gender, environment, the justice system, poverty, the political system, history, archaeology, anthropology, law, and many other fields that impact society. This makes social science research very important for the proper understanding of society and the appropriate collection of data and analysis to address social issues. Social scientists study society and the relationship of individuals within it. Their work can expose inequities and biases in society, from the workplace to education to healthcare. Unfortunately, the lack of political will to tackle critical social issues, and their politicization by some leaders, make the outstanding findings of scientists largely academic. That is until society demands it, sometimes in violent ways, as occurred in the wake of George Floyd's killing and caused the world to embrace the Black Lives Matter movement. I believe advocating for the government to address social issues is right down the alley of religious folks. If done more enthusiastically and purposefully, without injecting politics into it, it could lead to more legislation for social change for a better world.

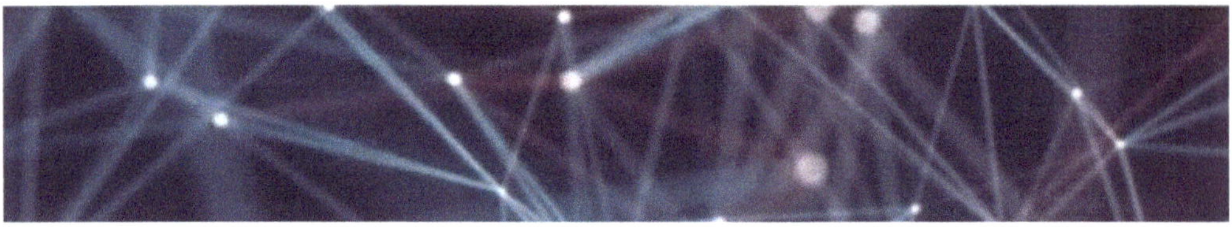

Certain global and moral questions are worth considering at the intersection of science, religion, and politics in society. Through science, we now understand the genetic under pinnings of numerous debilitating diseases (e.g., HIV/AIDS, Huntington's disease, cystic fibrosis, sickle cell anemia) that are simply inherited (conditioned by one gene).

These monogenic (single genes) dispositions to diseases are amenable to genetic engineering techniques such as CRISPR-cas9. The question is one of therapy versus enhancement. All efforts thus far have been geared towards using science for therapeutic purposes curing diseases when they occur. But what about preventing the disease from occurring in the first place? To do this will move scientific applications into the territory of enhancement, whereby the germline will be subjected to genetic manipulation. Some have asked whether we owe it to the future generation to use modern technology to correct genetic diseases at their source so that they are removed from human populations. That is, cystic fibrosis, for example, will be corrected such that it will not be inherited (parents cannot pass the disease to their children).

In a previous chapter, I indicated that genetic modification (germline and somatic) is routine in plant breeding (plant improvement or enhancement). Plant breeding is about making desired heritable changes in the germline. We owe the modern produce and animal products we purchase from the grocery store to plant and animal breeding. Using genetic modification technologies in plant breeding has been met with significant public outcry. The products of such efforts are called genetically modified organisms (GMOs) and contain alien genetic material in their genome.

Flagship products are the Bt and herbicide tolerant varieties in major crops like corn, potato, soybean, and cotton. Politicians have made decisions to ban GM products outright in some countries. Most corn and soybean produced in the US are GM.

On the animal side, one of the most widely accepted applications of genetic modification is insulin production. Commercial production of this essential drug is achieved through inserting the human gene for insulin into a microbe (yeast) so it can be expressed in large quantities in vats. I have yet to hear about the public protest of this application of genetic engineering. But the public has a right to be concerned about the modern biotechnologies we use in research, for humans are finite in their understanding and prediction of outcomes from applications of such technologies. The government should continue to police such enterprises to ensure the environment's health, safety, and well being are not jeopardized.

Much of the debate about protecting the environment centers around economics. Replacing fossil fuels in society will be good for the environment but bad for the industries associated with it and jobs that would be lost. And so, this has become a huge political fight between scientists and politicians. Clean energy is the way of the future. Politicians will have to accept this fact and implement policies that will help the transition between the old and new (e.g., retooling, creating new clean energy jobs, etc.).

Some Christian denominations have commissioned task forces to discuss some of these controversial social issues and declared their positions on them from the recommendations.

Some pronouncements (albeit dated) are as follows: "Human beings have an ability to do God like things: to exercise creativity, to direct and redirect processes of nature. But the warnings also imply that these powers may be used rashly, that it may be better for people to remember that they are creatures and not gods" (National Council of the Churches of Christ - 1980 Task Force report, Human Life and the New Genetics). Also, "The image of God, in which humanity is created, confers both power and responsibility to use power as God does: neither by coercion nor tyranny, but by love. Failure to accept limits by rejecting or ignoring accountability to God and interdependency with the whole of creation is the essence of sin" (United Methodist Church - 1992 Genetic Science Task Force report).

I believe I have discussed enough examples to demonstrate what happens when science, religion, and politics interplay in society. I stated elsewhere that psychologists generally believe all humans suffer from biases in what we think and do in society. The challenge for us is that even though we are unaware when we exercise these prejudices, the consequences are nonetheless real. Our holier than thou hypocrisies get in the way of seeing our world as it is. When we attempt to gather to discuss issues, especially those dear to our hearts, we talk at or past each other. Just like what I call a "Grand Canyon Conference." We agree to assemble, but in our minds, we are positioned at opposite sides of a canyon, with a deep chasm between us. We talk, literally to ourselves, for we cannot hear each other over the divide, be it ideological, religious, philosophical, or whatever else. We go back to our bases having gone through the motions without any resolution to problems. Consequently, the emotions are pent up until one day, a George Floyd caliber incident lights the fuse, forcing a violent outburst of frustration.

Politicians do what they do best, or is it worst, and capitalize on such situations and spin them for political advantage. This is when religion (Christianity) has an opportunity to step in, to bridge the divide by speaking truth in love, coupled with action for change. But alas, sometimes, you look out, but in vain, for the vocal disciples have abandoned their Master in His time of need. Sound familiar?

In closing, let me reiterate that the main message of this book is to state that God does not object to scientists interrogating nature, discovering natural laws, and applying their learned skills and techniques to nudge nature for the betterment of society. In fact, I believe it is all part of His divine plan for His creation.

But because we are finite in our understanding and knowledge, we have to be cautious in we approach our creative and innovative endeavors, for "we know not," as Scripture says. But one thing we know is that God is love, and He wants us to love Him above all else and love our neighbor as ourselves. We are commanded and expected to reachout to all people to invite them to come to Jesus. We can and should do better than we have and are doing now.

Thank God for STEM: Living by Faith and by Sight

A production of George Acquaah Publishing

A companion YouTube series for this book entitled, "Thank God for Science" is also available.

Georgeacguaahpublishing.com

In association with
Free Zone Productions

www.ingramcontent.com/pod-product-compliance
Ingram Content Group UK Ltd.
Pitfield, Milton Keynes, MK11 3LW, UK
UKHW061430110926
15112UKWH00025B/216

9 781958 518144